Pawel Delimata, Mikhail Ju. Moshkov, Andrzej Skowron,
and Zbigniew Suraj

Inhibitory Rules in Data Analysis

Studies in Computational Intelligence, Volume 163

Editor-in-Chief

Prof. Janusz Kacprzyk
Systems Research Institute
Polish Academy of Sciences
ul. Newelska 6
01-447 Warsaw
Poland
E-mail: kacprzyk@ibspan.waw.pl

Pawel Delimata
Mikhail Ju. Moshkov
Andrzej Skowron
Zbigniew Suraj

Inhibitory Rules in Data Analysis

A Rough Set Approach

 Springer

Pawel Delimata
Chair of Computer Science
University of Rzeszów
Rejtana 16A
35-310 Rzeszów
Poland
Email: pdelimata@wp.pl

Andrzej Skowron
Institute of Mathematics
Warsaw University
Banacha 2
02-097 Warsaw
Poland
Email: skowron@mimuw.edu.pl

Mikhail Ju. Moshkov
Institute of Computer Science
University of Silesia
Będzinska 39
41-200 Sosnowiec
Poland
Email: moshkov@us.edu.pl

Zbigniew Suraj
Chair of Computer Science
University of Rzeszów
Rejtana 16A
35-310 Rzeszów
Poland
Email: zsuraj@univ.rzeszow.pl

ISBN 978-3-540-85637-5   e-ISBN 978-3-540-85638-2

DOI 10.1007/978-3-540-85638-2

Studies in Computational Intelligence   ISSN 1860949X

Library of Congress Control Number: 2008933227

© 2009 Springer-Verlag Berlin Heidelberg

Typeset & Cover Design: Scientific Publishing Services Pvt. Ltd., Chennai, India.

Printed in acid-free paper

9 8 7 6 5 4 3 2 1

springer.com

To our families

Preface

This monograph is devoted to theoretical and experimental study of inhibitory decision and association rules. Inhibitory rules contain on the right-hand side a relation of the kind *"attribut $\neq$ value"*. The use of inhibitory rules instead of deterministic (standard) ones allows us to describe more completely information encoded in decision or information systems and to design classifiers of high quality.

The most important feature of this monograph is that it includes an advanced mathematical analysis of problems on inhibitory rules. We consider algorithms for construction of inhibitory rules, bounds on minimal complexity of inhibitory rules, and algorithms for construction of the set of all minimal inhibitory rules. We also discuss results of experiments with standard and lazy classifiers based on inhibitory rules. These results show that inhibitory decision and association rules can be used in data mining and knowledge discovery both for knowledge representation and for prediction. Inhibitory rules can be also used under the analysis and design of concurrent systems.

The results obtained in the monograph can be useful for researchers in such areas as machine learning, data mining and knowledge discovery, especially for those who are working in rough set theory, test theory, and logical analysis of data (LAD). The monograph can be used under the creation of courses for graduate students and for Ph.D. studies.

The authors of this book extend an expression of gratitude to Professor Janusz Kacprzyk, to Dr. Thomas Ditzinger and to the Studies in Computational Intelligence staff at Springer for their support in making this book possible.

Poland, 2008

Pawel Delimata
Mikhail Ju. Moshkov
Andrzej Skowron
Zbigniew Suraj

Contents

Introduction

The monograph is devoted to the study of *inhibitory* rules. In contrast to *deterministic* (standard) rules which have the relation *attribute = value* on the right-hand side, inhibitory rules have on the right-hand side the relation *attibute ≠ value*. For information systems [60, 63, 72], we consider inhibitory *association* rules which on the right-hand side can have an arbitrary attribute. For decision systems (decision tables) [60], we consider inhibitory *decision* rules containing on the right-hand side the decision attribute. It is worthwhile mentioning that in the rough set approach the decision rules are used for extension of approximations of concepts from given samples of objects on the whole universe of objects (see, e.g., [5, 6, 75]).

The use of inhibitory rules allows us to represent more knowledge encoded in information systems or decision tables. As a result, classifiers based on inhibitory rules are often more precise than the classifiers based on deterministic rules. Properties of inhibitory rules can be also used in the analysis and design of concurrent systems [58, 59, 61, 73, 84].

Let us discuss an example explaining the difference between deterministic and inhibitory decision rules. We consider the decision table T containing four rows (objects), two conditional attributes a_1 and a_2 and the decision attribute d:

$$T = \begin{array}{ccc|c} a_1 & a_2 & d \\ \hline 0 & 1 & 1 \\ 0 & 2 & 2 \\ 1 & 0 & 2 \\ 2 & 0 & 3 \end{array}$$

We study true and realizable decision rules for T. *True* means that the rule is true for each row of T. *Realizable* means that the left-hand side of the rule is true for at least one row of T. The rule

$$a_1 = 1 \rightarrow d = 2$$

is an example of true and realizable deterministic decision rule. Note that for any deterministic rule one can construct a set of inhibitory rules that is equivalent

P. Delimata et al.: Inhibitory Rules in Data Analysis, SCI 163, pp. 1–8.
springerlink.com   © Springer-Verlag Berlin Heidelberg 2009

to the deterministic rule. The considered deterministic rule is equivalent to the following set of true and realizable inhibitory rules:

$$a_1 = 1 \rightarrow d \neq 1 , \quad a_1 = 1 \rightarrow d \neq 3 .$$

Note that this set can be encoded by the following generalized rule:

$$a_1 = 1 \rightarrow d \neq 1 \wedge d \neq 3 ,$$

also denoted by

$$a_1 = 1 \rightarrow d \in V_d \setminus \{1, 3\} ,$$

where V_d is the value set of the decision d.

Based on deterministic or inhibitory decision rules, we can design classifiers corresponding to the table T. For any new object represented by values of conditional attributes a_1 and a_2, a classifier should generate a value of the decision attribute d for this object.

Let us consider the new object $o = (0, 0)$. For this object $a_1(o) = a_2(o) = 0$. It is clear that there is no true and realizable deterministic decision rule which is applicable to this object. Applicable to $(0, 0)$ means that the left-hand side of the rule is true for $(0, 0)$. However, there exist two true and realizable inhibitory decision rules which are applicable to $(0, 0)$:

$$a_1 = 0 \rightarrow d \neq 3 , \quad a_2 = 0 \rightarrow d \neq 1 .$$

The use of these rules allows us to conclude that the value of d for $(0, 0)$ is equal to 2.

The considered example shows that true and realizable inhibitory decision rules can represent more knowledge contained in decision tables than true and realizable deterministic decision rules. The same situation is with true and realizable inhibitory and deterministic association rules for information systems. Moreover, for association rules we prove the superiority of inhibitory rules over deterministic ones as a mathematical result.

Let $S = (U, A)$ be an information system [60, 63, 72], where U is a finite set of objects and A is a finite set of attributes (functions defined on U). We consider both deterministic and inhibitory association rules of the following form:

$$(a_1 = b_1) \wedge \ldots \wedge (a_t = b_t) \rightarrow a_{t+1} = b_{t+1} ,$$
$$(a_1 = b_1) \wedge \ldots \wedge (a_t = b_t) \rightarrow a_{t+1} \neq b_{t+1} ,$$

where $a_1, \ldots, a_{t+1}$ are attributes from A and $b_1, \ldots, b_{t+1}$ are values of these attributes. We consider only true and realizable rules. *True* means that the rule is true for any object from U. *Realizable* means that the left-hand side of the rule is true for at least one object from U. We identify objects from U and tuples of values of attributes from A on these objects[1]. Let V be the set of all tuples of

---

[1] i.e., we do not take into account the frequency with which tuples can appear [60] in data tables.

known values of attributes from A. In other words, V is the Cartesian product of ranges of attributes from A.

We say that the set U can be described by deterministic (inhibitory) rules if there exists a set Q of true and realizable deterministic (inhibitory) rules such that the set of objects from V, for which all rules from Q are true, coincides with U. In [73, 85], it was shown that there exist information systems $S = (U, A)$ such that the set U can not be described by deterministic rules. In Chap. 1, we show that for any information system $S = (U, A)$ the set U can be described by inhibitory rules.

In fact, in Chap. 1 we study the notion of *maximal consistent extension* of the set U for information system $S = (U, A)$ relative to a set R of all true for S rules of some kind. The maximal consistent extension of U relative to R is the set E of all objects from V for which each rule from R is true. The maximal consistent extensions play important role in investigations at the intersection of the rough set theory and the theory of concurrent systems [61, 73, 74, 85, 86]. As the set R, we consider (i) the set of all true inhibitory association rules, (ii) the set of all true and realizable inhibitory association rules, (iii) the set of all true deterministic association rules, and (iv) the set of all true and realizable deterministic association rules. In the first two cases, there are no nontrivial maximal consistent extensions E such that $E \neq U$. In the last two cases nontrivial maximal consistent extensions exist.

It means that the inhibitory association rules can express essentially more information encoded in information systems than the deterministic rules. This fact is a motivation for wider use of inhibitory rules in classification algorithms and for design of concurrent systems. There is an additional (intuitive) motivation for the use of inhibitory rules in classification algorithms: the inhibitory decision and association rules have much more chance to have larger support than the deterministic ones.

There are three ways to use inhibitory rules in classifiers: (i) lazy algorithms based on an information about the set of all inhibitory rules, (ii) standard classifiers based on a subset of inhibitory rules constructed by a heuristic, and (iii) standard classifiers based on the set of all minimal (irreducible) inhibitory rules.

In Chap. 2, we show that the approach (iii) is not feasible from computational point of view. We restrict our considerations to the class of k-valued information systems, i.e., information systems with attributes having values from $\{0, \ldots, k-1\}$, where $k \geq 2$. We show that, for any positive real number α, for almost all k-valued information systems S with m attributes and $n = \lfloor m^\alpha \rfloor$ objects the length of each minimal inhibitory rule is logarithmic relative to the number m. We also prove that for almost all k-valued information systems with m attributes and $n = \lfloor m^\alpha \rfloor$ objects the number of minimal inhibitory rules is at least $m^{(1/5)\alpha \log_k m}$ and at most $m^{2(1+\alpha)\log_k m + 7}$. Hence, from this result it follows that for almost all k-valued information systems with the polynomial number of objects in the number of attributes the number of minimal inhibitory rules is not polynomial in the number of attributes. Based on the obtained bounds we prove for k-valued information systems with m attributes and $n = \lfloor m^\alpha \rfloor$ objects that

- there is no algorithm which for almost all k-valued information systems constructs all minimal inhibitory rules and for such information systems has polynomial time complexity relative to the length of input,
- there exists an algorithm which for almost all k-valued information systems constructs all minimal inhibitory rules and for such information systems has polynomial time complexity relative to the sum of input length and the output length.

The obtained results have interesting interpretation for k-valued decision systems (decision tables). If for any considered information system S one attribute is selected as a decision attribute we obtain a k-valued decision system T. Minimal inhibitory rules of S with the selected decision attribute on their right-hand sides are called the minimal inhibitory decision rules of T. Such rules can be used for construction of classification algorithms (called classifiers, for short). Let us consider k-valued information systems S with m attributes and $n = \lfloor m^\alpha \rfloor$ objects. Then, for almost all such systems S all minimal inhibitory decision rules for each decision system obtained from S are relatively short (i.e., the length of each rule is logarithmic relative to the number m). However, for at least one decision system obtained from S the number of minimal inhibitory decision rules is at least $m^{(1/5)\alpha \log_k m - 1}$ (i.e., is not polynomial in the number of conditional attributes and the number of objects).

From these facts it follows that either it is necessary to use some heuristics for generating from a given information or decision system only a subset of "important" rules, or it is necessary to use lazy classification algorithms.

The heuristics for generating of inhibitory rules are considered in Chaps. 3 and 4. We restrict our consideration to inhibitory decision rules only. All obtained results are applicable to inhibitory association rules too. To this end we should fix an arbitrary attribute in the considered information system as the decision attribute, and study inhibitory decision rules for the obtained decision system.

In Chap. 3, we try to minimize the length of inhibitory decision rules (the number of conditions on the left-hand side). We consider not only exact but also partial (approximate) inhibitory rules: exact rules can be overfitted, i.e., dependent essentially on the noise or adjusted too much to the existing examples. If rules are considered as a way of knowledge representation [71], then instead of an exact rule with many attributes, it is more appropriate to work with a partial rule containing smaller number of attributes and having relatively high accuracy.

This idea is not new. For years, in rough set theory partial reducts and partial rules are studied intensively by J.G. Bazan, M.Ju. Moshkov, H.S. Nguyen, Z. Pawlak, M. Piliszczuk, A. Skowron, D. Ślęzak, J. Wróblewski, B. Zielosko and others (see., e.g., [2, 3, 4, 5, 26, 31, 32, 33, 39, 40, 42, 45, 47, 48, 55, 56, 62, 63, 65, 66, 78, 79, 80, 81, 82, 89, 94]). Approximate reducts are also investigated by W. Ziarko and M. Quafafou in the extensions of rough set model such as VPRS (variable precision rough sets) [93] and α-RST (alpha rough set theory) [67], respectively.

The theoretical studies of partial inhibitory decision rules is based on investigations of partial covers. There exists a simple reduction of the problem of construction of a cover with minimal cardinality to the problem of construction of an inhibitory decision rule with the minimal length. There also exists the opposite reduction which is simple too. The analogous situation is with partial covers and partial rules. This fact allows us to use various mathematical results obtained for the set cover problem by J. Cheriyan and R. Ravi [10], V. Chvátal [11], U. Feige [18], D.S. Johnson [23], R.M. Karp [24], M.J. Kearns [25], L. Lovász [28], M.Ju. Moshkov, M. Piliszczuk and B. Zielosko [41, 43, 44, 45, 47, 48], R.G. Nigmatullin [57], R. Raz and S. Safra [68], and P. Slavík [76, 77] for analysis of partial inhibitory rules.

In particular, we show that, under some natural assumptions on the class NP, the greedy algorithm is close to the best polynomial approximate algorithms for the minimization of the length of partial inhibitory decision rules. Based on an information received during the greedy algorithm work, it is possible to obtain nontrivial lower and upper bounds on the minimal length of partial inhibitory decision rules. For the most part of randomly generated binary decision tables, greedy algorithm constructs short partial inhibitory rules with relatively high accuracy. In particular, we obtain a result which confirms the following informal "0.5-hypothesis" for inhibitory decision rules: in the most part of cases, greedy algorithm chooses an attribute that separates at least one-half of unseparated rows that should be separated. Similar hypotheses for covers, reducts and deterministic decision rules were formulated and studied by M.Ju. Moshkov, M. Piliszczuk and B. Zielosko in [48].

In Chap. 4, we study the case, where each conditional attribute has its own weight, and we should minimize the total weight of attributes in partial rule. The weight of attribute characterizes time complexity, cost or risk (as in medical or technical diagnosis) of attribute value computation. The most part of results obtained in Chap. 3 is generalized to the case of arbitrary natural weights.

We generalize the standard greedy algorithm with weights, and consider a greedy algorithm with two thresholds. The first threshold specifies the exactness of constructed partial rule, and the second one is a parameter of the considered algorithm. In [45, 48] it is proved that for the most part of set cover problems for similar algorithm with two thresholds there exists a weight function and values of thresholds such that the weight of partial cover constructed by the greedy algorithm with two thresholds is less than the weight of partial cover constructed by the standard greedy algorithm. Based on the greedy algorithm with two thresholds we create new polynomial time approximate algorithms for minimization of weights of partial inhibitory rules. Such algorithms have some practical interest.

In Chaps. 5–7, we compare classifiers based on exact inhibitory and deterministic rules. In each of these chapters, we create two similar classifiers or two families of similar classifiers: the first one is based on deterministic rules, and the second one is based on inhibitory rules. Next, we consider results of experiments with real-life decision tables from [54]. The obtained results show that classifiers based on inhibitory rules are often better than the classifiers based on deterministic rules.

In Chap. 5, we study standard classifiers based on deterministic and inhibitory decision rules.

The first approach to construction of classifiers is the following: for a given decision table we construct for each row an exact deterministic decision rule using the greedy algorithm. The obtained system of rules jointly with simple procedure of voting can be considered as a classifier. A deterministic rule applicable to a given object is a vote "pro" the decision from the right-hand side of the rule.

The second approach to construction of classifiers is the following: for a given decision table we construct for each row and each decision, which is different from the decision attached to the considered row, an exact inhibitory decision rule using greedy algorithm. The obtained system of rules jointly with simple procedure of voting can be considered as a classifier. An inhibitory rule applicable to a given object is a vote "contra" the decision from the right-hand side of the rule.

In 46 experiments the error rate of the classifier based on inhibitory rules is less than the error rate of the classifier based on deterministic rules. In 7 experiments the error rate of the classifier based on inhibitory rules is greater than the error rate of the classifier based on deterministic rules. In 11 experiments the classifier based on inhibitory rules and the classifier based on deterministic rules have the same error rates.

In Chap. 6, we introduce two similar families of lazy classifiers based on exact deterministic and inhibitory association rules, respectively. Next, using these families we compare the "power" of inhibitory and deterministic rules experimentally. We consider the following classification problem: for a given decision table T and a new object v one should generate a value of the decision attribute on v using values of conditional attributes on v. To this end, we divide the decision table T into a number of information systems S_i, $i \in Dec(T)$, where $Dec(T)$ is the set of values of the decision attribute in T. For $i \in Dec(T)$, the information system S_i contains only objects (rows) of T with the value of the decision attribute equal to i.

For each information system S_i and a given object v, it is constructed (using polynomial-time algorithm) the so-called characteristic table filled by numbers 0 and 1. Rows of this table are labeled with objects from S_i and columns are labeled with attributes from S_i. For any object u from S_i and for any attribute a from S_i, the characteristic table contains 0 at the intersection of row labeled with u and column labeled with a if and only if there exists a rule which (i) is true for each object from S_i, (ii) is realizable for u (the left-hand side of this rule is true for u), (iii) is not true for v, and (iv) has the attribute a on the right-hand side.

Based on the characteristic table the decision on the "degree" to which v belongs to S_i is made for any i, and a decision i with the maximal "degree" is selected. We use different measures over characteristic tables for computing "degrees" to which v belongs to S_i. Let us consider two examples of such measures: the first measure is equal to the number of 1 in the characteristic table, and the second measure is equal to the number of rows in the characteristic table filled by 1 only.

If the characteristic table is created from deterministic rules, then we obtain a lazy classification algorithm based on deterministic association rules. If the characteristic table is created from inhibitory rules, then we obtain a lazy classification algorithm based on inhibitory association rules.

In 101 performed experiments the error rate of the classification algorithm based on inhibitory rules is less than the error rate of the classification algorithm based on deterministic rules. In 67 experiments the error rate of the algorithm based on inhibitory rules is greater than the error rate of the algorithm based on deterministic rules. In 72 experiments the algorithm based on inhibitory rules and the algorithm based on deterministic rules have the same error rates.

In Chap. 7, the same classification problem is considered: for a given decision table T and a new object v it is required to generate a value of the decision attribute d on v using values of conditional attributes on v. We compare two lazy classification algorithms based on deterministic and inhibitory decision rules of the forms

$$(a_1 = b_1) \wedge \ldots \wedge (a_t = b_t) \to d = i \, ,$$
$$(a_1 = b_1) \wedge \ldots \wedge (a_t = b_t) \to d \neq i \, ,$$

respectively, where $a_1, \ldots, a_t$ are conditional attributes, $b_1, \ldots, b_t$ are values of these attributes, d is the decision attribute and i is a value of d. By $Dec(T)$ we denote the set of values of the decision attribute d.

The first algorithm was created and studied by J.G. Bazan [2, 3, 4]. This algorithm is based on deterministic decision rules. For a new object v and each decision $i \in Dec(T)$ we find (using polynomial-time algorithm) the number $D(T, i, v)$ of objects u from the decision table T such that there exists a deterministic decision rule r satisfying the following conditions: (i) r is true for the decision table T, (ii) r is realizable for u and v, and (iii) r has the equality $d = i$ on the right-hand side. For the new object v we choose a decision $i \in Dec(T)$ for which the value $D(T, i, v)$ is maximal.

The second algorithm is based on inhibitory decision rules. For a new object v and each decision $i \in Dec(T)$ we find (using polynomial-time algorithm) the number $I(T, i, v)$ of objects u from the decision table T such that there exists an inhibitory decision rule r satisfying the following conditions: (i) r is true for the decision table T, (ii) r is realizable for u and v, and (iii) r has the relation $d \neq i$ on the right-hand side. For the new object v we choose a decision $i \in Dec(T)$ for which the value $I(T, i, v)$ is minimal.

In 29 performed experiments the error rate of the classification algorithm based on inhibitory rules is less than the error rate of the classification algorithm based on deterministic rules. In 11 experiments the error rate of the algorithm based on inhibitory rules is greater than the error rate of the algorithm based on deterministic rules. In 24 experiments the algorithm based on inhibitory rules and the algorithm based on deterministic rules have the same error rates.

Theoretical and experimental results presented in the monograph show that the inhibitory rules provide a powerful tool for representation of knowledge encoded in information and decision systems. Moreover, the inhibitory rules can be used in classification algorithms as well as deterministic rules. The inhibitory rules can be used also under the analysis and design of concurrent systems.

The results obtained in this monograph can be useful for researchers in such areas as machine learning, data mining and knowledge discovery, especially for those who are working in rough set theory and related theories such as test theory [9, 17, 83, 90, 91, 92] and logical analysis of data (LAD) [7, 12]. The monograph can be used under the creation of courses for graduate students and for Ph.D. studies.

The research has been partially supported by the grant N N516 368334 from Ministry of Science and Higher Education of the Republic of Poland and by the grant "Decision support – new generation systems" of Innovative Economy Operational Programme 2007-2013 (Priority Axis 1. Research and development of new technologies) managed by Ministry of Regional Development of the Republic of Poland.

1 Maximal Consistent Extensions of Information Systems

The idea of representation of concurrent system by information system is due to Z. Pawlak [61, 73, 74][1] In such a representation attributes are interpreted as local processes of a concurrent system, values of attributes – as states of local processes, and objects (tuples of values of attributes on objects) – as global states of the considered concurrent system.

The knowledge encoded in an information system can be represented by means of true rules which can be extracted from the information system. Besides "explicit" global states, corresponding to objects, the concurrent system generated by the considered information system can also have "hidden" global states, i.e., tuples of attribute values not belonging to a given information system but consistent with all the rules. Such "hidden" states can also be considered as realizable global states. This was a motivation for introducing in [73] maximal consistent extensions of information systems with both "explicit" and "hidden" global states. The maximal consistent extensions of information systems relative to the set of true and realizable deterministic rules were investigated in [70, 73, 85, 86].

In this chapter, we consider maximal consistent extensions of information systems relative to different kinds of rule sets. Let $S = (U, A)$ be an information system, where U is a finite set of objects and A is a finite set of attributes defined on U. We identify objects and tuples of values of attributes from A on these objects. Let V be the set of all possible objects – the Cartesian product of ranges of attributes from A, and let R be the set of all rules of a fixed kind which are true on U. The maximal consistent extension of U relative to R is the set E of all objects from V for which each rule from R is true. The maximal consistent extension of S relative to R can be represented by the information system $S^* = (E, A)$.

For different kinds of rule sets we discuss the following problems:

- the problem of existence of maximal consistent extensions E such that $E \neq U$,
- the problem of structure of E (in particular, the problem of cardinality of E and the problem of a distance between elements from $E \setminus U$ and U),

---

[1] Nowadays, discovery of process models from data becomes a hot topic under the name *process mining* (see, e.g., [8, 22, 30, 87]).

P. Delimata et al.: Inhibitory Rules in Data Analysis, SCI 163, pp. 9–29.
springerlink.com   © Springer-Verlag Berlin Heidelberg 2009

- the membership problem for E,
- the problem of construction of E in polynomial time,
- the problem of the minimal cardinality of a subset of R defining E,
- the problem of construction (in polynomial time) a subset of R defining E.

We study the following kinds of rule sets R: (i) the set of true and realizable inhibitory rules, (ii) the set of true inhibitory rules, (iii) the set of true deterministic rules, and (iv) the set of true and realizable deterministic rules.

Any deterministic rule has on the right-hand side a descriptor, i.e., an expression of the form $a(x) = v$, where a is an attribute and v is an attribute value, i.e., $v \in V_a$, where V_a is the range of a. In the case of inhibitory rules instead of descriptors on the right-hand sides of such rules there are negations of descriptors, i.e., expressions of the form $a(x) \neq v$. This means that the value of a on x belongs to $V_a \setminus \{v\}$. Let us recall that a rule is *true* if the rule is true for any object from U and a rule is *realizable* if the left-hand side of the rule is true for some object from U.

We show that in the cases of true and realizable inhibitory rules and true inhibitory rules there are no nontrivial maximal consistent extensions E (such that $E \neq U$), and there exist polynomial algorithms for construction of relatively small systems of rules describing exactly the set E.

In the case of true deterministic rules nontrivial maximal consistent extensions E exist, the structure of such extensions is simple, and cardinality of the set $E \setminus U$ is relatively small. Moreover, there exist polynomial algorithms for construction of the set E and a relatively small system of rules describing exactly the set E.

In the case of true realizable deterministic rules nontrivial maximal consistent extensions E exist [73, 85] but the cardinality of the set $E \setminus U$ can be very large. There is no polynomial algorithm for construction of the set E. It is an open problem if there exists a totally polynomial algorithm for construction of the set E (*totally polynomial* means polynomial in time, depending on the length of input and output). Also we do not know if there exist polynomial algorithms for construction of a system of rules describing exactly the set E. However, there exists a polynomial algorithm recognizing, for any given object from V, if this object belongs to E, or not [70].

The most important corollary from the obtained results is the following: the inhibitory rules can represent noticeably more knowledge encoded in information systems than the deterministic rules. Moreover, the support of inhibitory rules can be much larger than the deterministic ones. These facts are a motivation for the introduction of inhibitory rules into classification algorithms and into algorithms used for analysis and synthesis of concurrent systems.

This chapter is based on papers [49, 50, 52].

This chapter consists of six sections. In Sect. 1.1, we introduce the main notions. In Sects. 1.2–1.5, we study maximal consistent extensions relative to the set of true and realizable inhibitory rules, the set of true inhibitory rules, the set of true deterministic rules, and the set of true and realizable deterministic rules, respectively. Section 1.6 contains short conclusions.

1.1   Main Notions

Let $S = (U, A)$ be an information system [60], where $U = \{u_1, \ldots, u_n\}$ is a set of objects and $A = \{a_1, \ldots, a_m\}$ is a set of attributes (functions defined on U). We assume that for any two different numbers $i, j \in \{1, \ldots, n\}$ tuples $(a_1(u_i), \ldots, a_m(u_i))$ and $(a_1(u_j), \ldots, a_m(u_j))$ are different. Hence, we may identify $u_i \in U$ and the corresponding tuple $(a_1(u_i), \ldots, a_m(u_i))$ for $i = 1, \ldots, n$.

For $j = 1, \ldots, m$, we denote by V_{a_j} the set $\{a_j(u_i) : u_i \in U\}$. We assume that $|V_{a_j}| \geq 2$ for $j = 1, \ldots, m$. Moreover, let $V(S) = V_{a_1} \times \ldots \times V_{a_m}$. We assume that $U \neq V(S)$.

We consider extensions U^* of the set U such that $U \subseteq U^* \subseteq V(S)$. We also assume that, for any $a_j \in A$ and any $u \in V(S)$, the value $a_j(u)$ is equal to j-th component of u.

Let us consider a rule

$$(a_{j_1}(x) = b_1) \wedge \ldots \wedge (a_{j_{t-1}}(x) = b_{t-1}) \rightarrow a_{j_t}(x) \neq b_t , \qquad (1.1)$$

where $t \geq 1$, $a_{j_1}, \ldots, a_{j_t} \in A$, $b_1 \in V_{a_{j_1}}, \ldots, b_t \in V_{a_{j_t}}$, and numbers $j_1, \ldots, j_t$ are pairwise different. Such rules are called *inhibitory association* rules. Later we will omit the word "association". The rule (1.1) is *true for an object* $u \in V(S)$ if there exists $l \in \{1, \ldots, t-1\}$ such that $a_{j_l}(u) \neq b_l$, or $a_{j_t}(u) \neq b_t$. The rule (1.1) is *true* if it is true for any object from U. The rule (1.1) is *realizable* if there exists an object $u_i \in U$ such that $a_{j_1}(u_i) = b_1, \ldots, a_{j_{t-1}}(u_i) = b_{t-1}$.

Let us consider a rule

$$(a_{j_1}(x) = b_1) \wedge \ldots \wedge (a_{j_{t-1}}(x) = b_{t-1}) \rightarrow a_{j_t}(x) = b_t , \qquad (1.2)$$

where $t \geq 1$, $a_{j_1}, \ldots, a_{j_t} \in A$, $b_1 \in V_{a_{j_1}}, \ldots, b_t \in V_{a_{j_t}}$, and numbers $j_1, \ldots, j_t$ are pairwise different. Such rules are called *deterministic association* rules. Later we will omit the word "association". The rule (1.2) is *true for an object* $u \in V(S)$ if there exists $l \in \{1, \ldots, t-1\}$ such that $a_{j_l}(u) \neq b_l$, or $a_{j_t}(u) = b_t$. The rule (1.2) is *true* if it is true for any object from U. The rule (1.2) is *realizable* if there exists an object $u_i \in U$ such that $a_{j_1}(u_i) = b_1, \ldots, a_{j_{t-1}}(u_i) = b_{t-1}$.

Note that later we consider only information systems $S = (U, A)$, where $A = \{a_1, \ldots, a_m\}$, $|V_{a_j}| \geq 2$ for $j = 1, \ldots, m$, $(a_1(u), \ldots, a_m(u)) \neq (a_1(v), \ldots, a_m(v))$ for any $u, v \in U$, $u \neq v$, and $U \neq V(S)$.

1.2   True Realizable Inhibitory Rules

Let S be an information system. By $Rul^1(S)$ we denote the set of all inhibitory rules which are true and realizable in S. By $Ext^1(S)$ we denote the set of all objects $u \in V(S)$ such that each rule from $Rul^1(S)$ is true for u. The set $Ext^1(S)$ is called the *maximal consistent extension of U relative to the set of rules $Rul^1(S)$*.

1.2.1   Separating Sets of Attributes

A set of attributes $B \subseteq A$ is called *separating* U from $V(S) \setminus U$ if for any two objects $u \in U$ and $v \in V(S) \setminus U$ there exists an attribute $a_j \in B$ such that $a_j(u) \neq a_j(v)$, i.e., tuples u and v are different on j-th component. In this subsection, we describe all separating sets of attributes.

A separating set is called *irreducible* if its each proper subset is not a separating set. It is clear that the set of all irreducible separating sets coincides with the set of all reducts for the decision system $D = (V(S), A, d)$ [60, 63, 64, 65], where for any $u \in V(S)$,

$$d(u) = \begin{cases} 1, \text{ if } u \in U , \\ 0, \text{ if } u \notin U . \end{cases}$$

We show that the core for this decision system is a reduct. This means that D has exactly one reduct coinciding with the core.

By $C(S)$ we denote the set of all attributes $a_j \in A$ such that there exist two objects $u \in U$ and $v \in V(S) \setminus U$, which are different only on j-th component.

It is clear that $C(S)$ is the core for D, and $C(S)$ is a subset of each reduct for D.

Proposition 1.1. *The set $C(S)$ is a reduct for the decision system*

$$D = (V(S), A, d) .$$

Proof. Let us consider two objects $u \in U$ and $v \in V(S) \setminus U$. We will show that these objects are different on an attribute from $C(S)$. Let us assume that u and v are different on p components $j_1, \ldots, j_p$. Then, there exists a sequence $u_1, \ldots, u_{p+1}$ of objects from $V(S)$ such that $u = u_1$, $v = u_{p+1}$ and, for $i = 1, \ldots, p$, the objects u_i and u_{i+1} are different only on the component with number j_i. Since $u_1 \in U$ and $u_{p+1} \in V(S) \setminus U$, there exists $i \in \{1, \ldots, p\}$ such that $u_i \in U$ and $u_{i+1} \in V(S) \setminus U$. Therefore, $a_{j_i} \in C(S)$. It is clear that u and v are different on the attribute a_{j_i}. Thus, $C(S)$ is a reduct for D. $\square$

From Proposition 1.1 it follows that $C(S)$ is the unique reduct for the decision system D. Thus, a set $B \subseteq A$ is a separating set if and only if $C(S) \subseteq B$. One can show that $C(S) \neq \emptyset$ if and only if $U \neq V(S)$.

We now describe a polynomial algorithm $\mathcal{A}_1$ for construction of the set $C(S)$.

Remark 1.2. Reducts for the information system $S = (U, A)$ and the reduct for the decision system D can be very different. Let, for example, $m \geq 2$, $A = \{a_1, \ldots, a_m\}$ and $U = \{(0, 0, \ldots, 0), (1, 0, \ldots, 0), (2, 1, \ldots, 1)\}$. Then $C(S) = \{a_1, \ldots, a_m\}$, and S has only one reduct $\{a_1\}$.

Proposition 1.3 states that the set U cannot be exactly described by any system of true inhibitory rules in which some attributes from $C(S)$ do not occur.

Proposition 1.3. *Let Q be a set of true inhibitory rules, the set of objects from $V(S)$, for which any rule from Q it true, coincide with U, and let B – the set of attributes from A contained in rules from Q. Then $C(S) \subseteq B$.*

Algorithm 1.1. Algorithm $\mathcal{A}_1$ for construction $C(S)$

Input   : Information system $S = (U, A)$, where $A = \{a_1, \ldots, a_m\}$.
Output: Set $C(S)$.
$C(S) \longleftarrow \emptyset$;
for $(b_1, \ldots, b_m) \in U$; $j = 1, \ldots m$; $c \in V_{a_j} \setminus \{b_j\}$ **do**
    if $(b_1, \ldots, b_{j-1}, c, b_{j+1}, \ldots, b_m) \in V(S) \setminus U$ **then**
        $C(S) \longleftarrow C(S) \cup \{a_j\}$;
    end
end

Proof. Let us assume the contrary: B does not contain an attribute $a_j \in C(S)$. Since $a_j \in C(S)$, there exists a pair of objects $u \in U$ and $v \in V(S) \setminus U$, which are different only on the component with number j. Let us consider a rule (1.1) from Q which is not true for the object v. Since this rule does not contain the attribute a_j, the considered rule is not true for u which is impossible. □

1.2.2   Description of the Set U

In this subsection, we show that U can be described by a set of true realizable inhibitory rules with attributes from $C(S)$. It means that $Ext^1(S) = U$.

Remind that $U \neq V(S)$. Let F be a tuple of pairwise different attributes from A containing all attributes from $C(S)$. Let, for the definiteness, $F = (a_1, \ldots, a_p)$. We now describe an algorithm $\mathcal{A}_2$ constructing a decision tree $\Gamma(S, F)$ for the decision system D which recognizes the value of the decision attribute d using only values of attributes from F.

Taking into account that $C(S) \subseteq \{a_1, \ldots, a_p\}$ one can show that $\Gamma = \Gamma(S, F)$ solves the problem of membership to U: for each $u \in U$ the work of Γ ends in a terminal node labeled with 1, and for each $v \in V(S) \setminus U$ the work of Γ ends in a terminal node labeled with 0.

Let us evaluate the number of nodes in the tree Γ.

Lemma 1.4. *The decision tree $\Gamma = \Gamma(S, F)$ has at most $pn \leq mn$ nonterminal nodes, at most $pnk' \leq mnk$ terminal nodes, and at most $pn(k'-1) \leq mn(k-1)$ terminal nodes labeled with 0, where p is the number of attributes in F, $m = |A|$, $n = |U|$, $k = \max\{|V_{a_j}| : a_j \in A\}$ and $k' = \max\{|V_{a_j}| : a_j \in F\}$.*

Proof. It is clear that for each nonterminal node there exists an object from U for which the computation passes through this node. For each object from U the computation passes through at most p nonterminal nodes. Therefore, there are at most $pn \leq mn$ nonterminal nodes in Γ.

It is clear that the number of terminal nodes is at most the number of nonterminal nodes multiplied by k'. Therefore, the number of terminal nodes in Γ is at most $pnk' \leq mnk$.

From the description of the algorithm $\mathcal{A}_2$ it follows that for each nonterminal node there exist at most $k' - 1$ terminal nodes, where each is connected with the

---

Algorithm 1.2. Algorithm $\mathcal{A}_2$ for construction $\Gamma(S, F)$

Input : Information system $S = (U, A)$, and tuple $F = (a_1, \ldots, a_p)$ of pairwise different attributes from A containing all attributes from $C(S)$, where $V_{a_j} = \{c_{j1}, \ldots, c_{jk_j}\}$ for $j = 1, \ldots, p$,

Output: Decision tree $\Gamma(S, F)$.

construct the tree G which has only one node v_0, and this node is not marked as treated;

while G *contains nodes that are not marked as treated* **do**

    select a node w in G which is not marked as treated;

    if $w = v_0$ **then**

        $t \longleftarrow 0$;

        $W \longleftarrow V(S)$;

    end

    if $w \neq v_0$ **then**

        find the path $w_1, q_1, \ldots, w_t, q_t, w$ from the root $w_1 = v_0$ of G to the node w; nodes $w_1, \ldots, w_t$ are labeled with attributes $a_1, \ldots, a_t$; let edges $q_1, \ldots, q_t$ be labeled with values $b_1, \ldots, b_t$; construct the set W of solutions from $V(S)$ of the equation system $\{a_1(x) = b_1, \ldots, a_t(x) = b_t\}$;

    end

    if $W \subseteq U$ **then**

        mark the node w by the number 1, and mark the node w as treated;

    end

    if $W \subseteq V(S) \setminus U$ **then**

        mark the node w by the number 0, and mark the node w as treated;

    end

    if $W \cap U \neq \emptyset$ *and* $W \cap (V(S) \setminus U) \neq \emptyset$ **then**

        add to the tree G nodes $v_1, \ldots, v_{k_{t+1}}$ and edges $e_1, \ldots, e_{k_{t+1}}$ issuing from w and entering $v_1, \ldots, v_{k_{t+1}}$ respectively; mark the node w by the attribute a_{t+1}; for $i = 1, \ldots, k_{t+1}$, mark the edge e_i by the value c_{t+1i}; mark the node w as treated; (since $C(S)$ is a reduct for D and $C(S) \subseteq \{a_1, \ldots, a_p\}$, we have $t + 1 \leq p$)

    end

end

$\Gamma(S, F) \longleftarrow G$;

---

considered node by an edge and is labeled with 0. Therefore, the number of terminal nodes in Γ which are labeled with 0 is at most $pn(k' - 1) \leq mn(k - 1)$. $\square$

Using Lemma 1.4 one can show that the algorithm $\mathcal{A}_2$ has polynomial time complexity.

We now describe an algorithm $\mathcal{A}_3$ which transforms the decision tree $\Gamma = \Gamma(S, F)$ into a set of equation systems $Eq(\Gamma)$. Let ρ be a path from the root to a terminal node of Γ, $\rho = v_1, e_1, \ldots, v_t, e_t, v_{t+1}$, nodes $v_1, \ldots, v_t$ be labeled with attributes $a_1, \ldots, a_t$, and edges $e_1, \ldots, e_t$ be labeled with values $b_1, \ldots, b_t$. By $ES(\rho)$ we denote the equation system $\{a_1(x) = b_1, \ldots, a_t(x) = b_t\}$.

It is clear that the considered algorithm has polynomial time complexity.

Algorithm 1.3. Algorithm $\mathcal{A}_3$ for construction $Eq(\Gamma)$

Input : Decision tree $\Gamma = \Gamma(S, F)$.
Output: Set of equation systems $Eq(\Gamma)$.
$Eq(\Gamma) \longleftarrow \emptyset$;
for *each path ρ from the root to a terminal node of Γ labeled with 0* **do**
| add the equation system $ES(\rho)$ to $Eq(\Gamma)$;
end

Lemma 1.5. *The set of equation systems $Eq(\Gamma) = Eq(\Gamma(S, F))$ has the following properties:*

1. *For each equation system from $Eq(\Gamma)$ any solution of this system from $V(S)$ belongs to $V(S) \setminus U$.*
2. *For each object $u \in V(S) \setminus U$ there exists an equation system from $Eq(\Gamma)$ such that u is a solution of this system.*
3. *In each equation system from $Eq(\Gamma)$ the number of equations is at least 2.*
4. *$|Eq(\Gamma)| \leq pn(k'-1) \leq mn(k-1)$, where p is the number of attributes in F and $k' = \max\{|V_{a_j}| : a_j \in F\}$.*

Proof. From the fact that Γ solves the problem of membership to U it follows that $Eq(\Gamma)$ has the first two properties. Let us assume that $Eq(\Gamma)$ contains a system of equations with one equation only. Then this system has a solution from U which is impossible. It is clear that $|Eq(\Gamma)|$ is equal to the number of terminal nodes in Γ which are labeled with 0. From Lemma 1.4 it follows that the considered number is at most $pn(k'-1) \leq mn(k-1)$. $\qquad\square$

Algorithm 1.4. Algorithm $\mathcal{A}_4$ for construction $R^1(\Gamma)$

Input : Set of equation systems $Eq(\Gamma)$.
Output: Set $R^1(\Gamma)$ of inhibitory rules.
$R^1(\Gamma) \longleftarrow \emptyset$;
for *each equation system $\{a_1(x) = b_1, \ldots, a_t(x) = b_t\}$ from $Eq(\Gamma)$* **do**
| add rule $(a_1(x) = b_1) \wedge \ldots \wedge (a_{t-1}(x) = b_{t-1}) \rightarrow a_t(x) \neq b_t$ to $R^1(\Gamma)$;
end

We now consider a polynomial algorithm $\mathcal{A}_4$ which transforms the set of equation systems $Eq(\Gamma)$ into a set $R^1(\Gamma)$ of inhibitory rules.

Lemma 1.6. *The set of rules $R^1(\Gamma) = R^1(\Gamma(S, F))$ has the following properties:*

1. *$R^1(\Gamma) \subseteq Rul^1(S)$.*
2. *Rules from $R^1(\Gamma)$ use only attributes from F.*
3. *The set of objects from $V(S)$, for which any rule from $R^1(\Gamma)$ it true, coincides with U.*
4. *$|R^1(\Gamma)| \leq pn(k'-1) \leq mn(k-1)$, where p is the number of attributes in F and $k' = \max\{|V_{a_j}| : a_j \in F\}$.*

Proof. From the descriptions of the decision tree Γ and the set $Eq(\Gamma)$ it follows that any equation system

$$\{a_1(x) = b_1, \ldots, a_t(x) = b_t\} \tag{1.3}$$

from $Eq(\Gamma)$ has no solutions from U, but the system obtained from (1.3) by removal of the last equation has a solution from U. Therefore, $R^1(\Gamma) \subseteq Rul^1(S)$. Hence, each rule from $R^1(\Gamma)$ is true for each element from U. From Lemma 1.5 it follows that for each element from $V(S) \setminus U$ there exists a rule from $R^1(\Gamma)$ which is not true for this element. Thus, the set of objects from $V(S)$, for which any rule from $R^1(\Gamma)$ is true, coincides with U. The inequalities $|R^1(\Gamma)| \le pn(k'-1) \le mn(k-1)$ follow from Lemma 1.5. It is clear that rules from $R^1(\Gamma)$ use only attributes from F. $\qquad\square$

Theorem 1.7. *There exists a polynomial algorithm constructing, for a given information system $S = (U, A)$, a set R of true realizable inhibitory rules with the following properties:*

1. *Rules from R use only attributes from $C(S)$.*
2. *The set of all objects from $V(S)$, for which any rule from R is true, coincides with U.*
3. *$|R| \le pn(k'-1) \le mn(k-1)$, where $p = |C(S)|$, $n = |U|$, $m = |A|$, $k' = \max\{|V_{a_j}| : a_j \in C(S)\}$, and $k = \max\{|V_{a_j}| : a_j \in A\}$.*

Proof. The considered algorithm is the following. Using algorithm $\mathcal{A}_1$ we construct the unique irreducible separating set $C(S)$ which is the core and unique reduct for the decision system $D = (V(S), A, d)$ corresponding to S. Let $C(S) = \{a_{j_1}, \ldots, a_{j_p}\}$. Set $F = (a_{j_1}, \ldots, a_{j_p})$. Using algorithm $\mathcal{A}_2$ we construct the decision tree $\Gamma = \Gamma(S, F)$. Using algorithms $\mathcal{A}_3$ and $\mathcal{A}_4$ we construct the set of equation systems $Eq(\Gamma)$ and the set of rules $R^1(\Gamma)$. Set $R = R^1(\Gamma)$.

It is clear that the considered algorithm has polynomial time complexity. From Lemma 1.6 it follows that the system of rules R has properties formulated in the theorem. $\qquad\square$

Corollary 1.8. *For any information system $S = (U, A)$ the following equality holds:*

$$Ext^1(S) = U .$$

Using Corollary 1.8 we conclude that, for true realizable inhibitory rules, problems related to membership to $Ext^1(S)$, evaluation of $|Ext^1(S)|$, structure of $Ext^1(S)$ and constriction of $Ext^1(S)$ have trivial solutions.

We now show that the bound $|R| \le mn(k-1)$ from Theorem 1.7 can not be improved essentially, in general case.

Let m, r and k be natural numbers such that $r < m$ and $k \ge 2$. Let $u \in \{0, 1, \ldots, k-1\}^m$. Let us denote the number of components in u equal to 0 by $l_0(u)$, and let

$$U(m, r, k) = \{u : u \in \{0, 1, \ldots, k-1\}^m, l_0(u) = m - r\} .$$

It is clear that $|U(m, r, k)| = C_m^r(k-1)^r$. By $A(m)$ we denote the set of attributes $\{a_1, \ldots, a_m\}$ such that, for any $a_j \in A$ and any $u \in \{0, 1, \ldots, k-1\}^m$, the value $a_j(u)$ is equal to the j-th component of u.

Proposition 1.9. *Let m, r and k be natural numbers such that $r < m$ and $k \geq 2$, $S = (U(m, r, k), A(m))$, and Q be a set of true inhibitory rules such that the set of objects from $V(S)$, for which each rule from Q is true, coincides with $U(m, r, k)$. Then*
$$|Q| \geq C_m^{r+1}(k-1)^{r+1} .$$

Proof. Let $v \in U(m, r+1, k)$. Then Q contains a rule (1.1) which is not true for v. Since (1.1) is true for all objects from $U(m, r, k)$ and not true for v, the system of equations
$$\{a_{j_1}(x) = b_1, \ldots, a_{j_t}(x) = b_t\} \tag{1.4}$$
has no solutions from $U(m, r, k)$, but v is a solution of (1.4).

One can show that (1.4) has no solutions from $U(m, r, k)$ if and only if among numbers $b_1, \ldots, b_t$ there are at least $m-r+1$ numbers 0, or at least $r+1$ numbers, which are not equal to 0. Since v is a solution of (1.4), among $b_1, \ldots, b_t$ exactly $r+1$ numbers are not equal to 0. Let, for the definiteness, $b_1, \ldots, b_{r+1}$ be not equal to 0. Then (1.4) is equal to

$$\{a_{j_1}(x) = b_1, \ldots, a_{j_{r+1}}(x) = b_{r+1}, a_{j_{r+2}}(x) = 0, \ldots, a_{j_t}(x) = 0\} ,$$

where $j_1, \ldots, j_{r+1}$ are numbers of components in which v has values which are not equal to 0, and $b_1, \ldots, b_{r+1}$ are components of v with numbers $j_1, \ldots, j_{r+1}$.

Let v_1 and v_2 be different objects from $U(m, r+1, k)$ and rul_1, rul_2 be rules from Q such that rul_1 is not true for v_1 and rul_2 is not true for v_2. From the obtained results it follows that the rules rul_1 and rul_2 are different. Therefore, $|Q| \geq C_m^{r+1}(k-1)^{r+1}$. □

Let m, r and k be natural numbers such that $r < m$ and $k \geq 2$, and let $S = (U(m, r, k), A(m))$. It is clear that $|A(m)| = m$, $|U(m, r, k)| = C_m^r(k-1)^r$ and $\max\{|V_{a_j}| : a_j \in A(m)\} = k$. The upper bound on the minimal number of rules $mn(k-1)$ from Theorem 1.7 is equal here to

$$mC_m^r(k-1)^r(k-1) .$$

Let Q be a set of true inhibitory rules such that the set of objects from $V(S)$, for which each rule from Q is true, coincides with $U(m, r, k)$. From Proposition 1.9 it follows that

$$|Q| \geq C_m^{r+1}(k-1)^{r+1} = \frac{m-r}{r+1} C_m^r(k-1)^r(k-1) .$$

One can see that for low values of r the obtained lower bound is not very far from the upper one.

1.3 True Inhibitory Rules

Let S be an information system. By $Rul^2(S)$ we denote the set of inhibitory and true rules. By $Ext^2(S)$ we denote the set of objects from $V(S)$ for which each rule from $Rul^2(S)$ is true. The set $Ext^2(S)$ is called *maximal consistent extension of U relative to the set of rules $Rul^2(S)$*.

From Proposition 1.3 it follows that there are no systems of true inhibitory rules which describe exactly the set U and do not use at least one attribute from $C(S)$.

Next statement (which follows immediately from Theorem 1.7) shows that using true inhibitory rules with attributes from $C(S)$ it is possible to describe exactly the set U.

Theorem 1.10. *There exists a polynomial algorithm constructing, for a given information system $S = (U, A)$, a set R of true inhibitory rules with the following properties:*

1. *Rules from R use only attributes from $C(S)$.*
2. *The set of all objects from $V(S)$, for which any rule from R is true, coincides with U.*
3. $|R| \leq pn(k' - 1) \leq mn(k - 1)$, *where* $p = |C(S)|$, $n = |U|$, $m = |A|$, $k' = \max\left\{\left|V_{a_j}\right| : a_j \in C(S)\right\}$, *and* $k = \max\left\{\left|V_{a_j}\right| : a_j \in A\right\}$.

Corollary 1.11. *For any information system $S = (U, A)$ the following equality holds:*
$$Ext^2(S) = U .$$

Using Corollary 1.11 we conclude that in the case of true inhibitory rules problems related to the membership to $Ext^2(S)$, evaluation of $\left|Ext^2(S)\right|$, structure of $Ext^2(S)$ and constriction of $Ext^2(S)$ have trivial solutions.

From Proposition 1.9 it follows that the bound $|R| \leq mn(k-1)$ from Theorem 1.10 can not be improved essentially, in general case.

1.4 True Deterministic Rules

Let S be an information system. By $Rul^3(S)$ we denote the set of all deterministic and true rules. By $Ext^3(S)$ we denote the set of objects from $V(S)$ for which each rule from $Rul^3(S)$ is true. The set $Ext^3(S)$ is called *maximal consistent extension of U relative to the set of rules $Rul^3(S)$*.

1.4.1 Information Systems with Binary Attribute

An attribute $a_j \in A$ is called *binary* if $\left|V_{a_j}\right| = 2$. Let us assume that the information system S contains a binary attribute a_{j_0}. Let, for the definiteness, $j_0 = 1$ and $C(S) \cup \{a_1\} = \{a_1, \ldots, a_p\}$. Let us denote $F = (a_1, \ldots, a_p)$ and $\Gamma = \Gamma(S, F)$. We now consider a polynomial algorithm $\mathcal{A}_5$ which transforms the set of equation systems $Eq(\Gamma)$ into a set $R^2(\Gamma)$ of deterministic rules.

Algorithm 1.5. Algorithm $\mathcal{A}_5$ for construction $R^2(\Gamma)$

Input : Set of equation systems $Eq(\Gamma)$.
Output: Set $R^2(\Gamma)$ of deterministic rules.
$R^2(\Gamma) \longleftarrow \emptyset$;
for *each equation system* $\{a_1(x) = b_1, \ldots, a_t(x) = b_t\}$ *from* $Eq(\Gamma)$ **do**
$\quad\mid\quad$ choose $b_1' \in V_{a_1} \setminus \{b_1\}$;
$\quad\mid\quad$ add rule $(a_2(x) = b_2) \wedge \ldots \wedge (a_t(x) = b_t) \rightarrow a_1(x) = b_1'$ to $R^2(\Gamma)$;
end

Lemma 1.12. *The set of rules* $R^2(\Gamma) = R^2(\Gamma(S, F))$ *has the following properties:*

1. $R^2(\Gamma) \subseteq Rul^3(S)$.
2. *Rules from* $R^2(\Gamma)$ *use only attributes from* F.
3. *The set of objects from* $V(S)$, *for which any rule from* $R^2(\Gamma)$ *it true, coincides with* U.
4. $|R^2(\Gamma)| \leq pn(k'-1) \leq mn(k-1)$, *where* p *is the number of attributes in* F, $k' = \max\{|V_{a_j}| : a_j \in F\}$, *and* $k = \max\{|V_{a_j}| : a_j \in A\}$.

Proof. From the definitions of the decision tree Γ and the set $Eq(\Gamma)$ it follows that any equation system (1.3) from $Eq(\Gamma)$ has no solutions from U. Therefore, $R^2(\Gamma) \subseteq Rul^3(S)$. Hence, each rule from $R^2(\Gamma)$ is true for each element from U. From Lemma 1.5 it follows that for each element from $V(S) \setminus U$ there exists a rule from $R^2(\Gamma)$ which is not true for this element. Thus, the set of objects from $V(S)$, for which any rule from $R^2(\Gamma)$ it true, coincides with U. The inequalities $|R^2(\Gamma)| \leq pn(k'-1) \leq mn(k-1)$ follow from Lemma 1.5. It is clear that rules from $R^2(\Gamma)$ use only attributes from F. $\qquad\square$

Theorem 1.13. *There exists a polynomial algorithm which, for a given information system* $S = (U, A)$ *with a binary attribute* a_{j_0}, *constructs a set* R *of true deterministic rules with the following properties:*

1. *Rules from* R *use only attributes from* $C(S) \cup \{a_{j_0}\}$.
2. *The set of objects from* $V(S)$, *for which any rule from* R *is true, coincides with* U.
3. $|R| \leq pn(k'-1) \leq mn(k-1)$, *where* $p = |C(S) \cup \{a_{j_0}\}|$, $n = |U|$, $k' = \max\{|V_{a_j}| : a_j \in C(S) \cup \{a_{j_0}\}\}$, $m = |A|$, *and* $k = \max\{|V_{a_j}| : a_j \in A\}$.

Proof. The considered algorithm is the following. Using algorithm $\mathcal{A}_1$ we construct the unique irreducible separating set $C(S)$ which is the core and unique reduct for the decision system $D = (V(S), A, d)$ corresponding to S. Let $C(S) \cup \{a_{j_0}\} = \{a_{j_0}, \ldots, a_{j_{p-1}}\}$. Set $F = (a_{j_0}, \ldots, a_{j_{p-1}})$. Using algorithm $\mathcal{A}_2$ we construct the decision tree $\Gamma = \Gamma(S, F)$. Using algorithms $\mathcal{A}_3$ and $\mathcal{A}_5$ we construct the set of equation systems $Eq(\Gamma)$ and the set of deterministic rules $R^2(\Gamma)$. Set $R = R^2(\Gamma)$.

It is clear that the considered algorithm has polynomial time complexity. From Lemma 1.12 it follows that the system of rules R has properties formulated in the theorem. □

Corollary 1.14. *For any information system $S = (U, A)$ with a binary attribute, the equality $Ext^3(S) = U$ holds.*

Using Corollary 1.14 we conclude that in the case of true deterministic rules problems related to the membership to $Ext^3(S)$, evaluation of $|Ext^3(S)|$, structure of $Ext^3(S)$ and construction of $Ext^3(S)$ have trivial solutions, if S contains a binary attribute.

Let us assume that S does not contain binary attributes. We show that in general case it is impossible to describe U exactly using true deterministic rules. However, there exist polynomial algorithms constructing the set $Ext^3(S)$ and a set of true deterministic rules describing exactly the set $Ext^3(S)$.

1.4.2   Structure of $Ext^3(S)$

By $I(S)$ we denote the set of all objects $u = (b_1, \ldots, b_m)$ from $V(S) \setminus U$ such that the system of equations

$$\{a_1(x) = b_1, \ldots, a_m(x) = b_m\} \tag{1.5}$$

has no solutions from U, but for each $j \in \{1, \ldots, m\}$ the system of equations

$$\{a_1(x) = b_1, \ldots, a_{j-1}(x) = b_{j-1}, a_{j+1}(x) = b_{j+1}, \ldots, a_m(x) = b_m\} \tag{1.6}$$

has two solutions $u_1 = (c_1, \ldots, c_m)$ and $u_2 = (d_1, \ldots, d_m)$ from U such that $c_j \neq b_j$, $d_j \neq b_j$ and $c_j \neq d_j$.

It is clear that if S has a binary attribute, then $I(S) = \emptyset$.

Theorem 1.15. *For any information system $S = (U, A)$, the following equality holds:*

$$Ext^3(S) = U \cup I(S) .$$

Proof. Each object from U belongs to $Ext^3(S)$. Let $u = (b_1, \ldots, b_m) \in I(S)$. We now show that u belongs to $Ext^3(S)$. Let us assume the contrary: a rule

$$(a_{j_1}(x) = e_1) \wedge \ldots \wedge (a_{j_{t-1}}(x) = e_{t-1}) \rightarrow a_{j_t}(x) = e_t$$

from $Rul^3(S)$ exists which is not true for u. It is clear that $e_1 = b_{j_1}, \ldots, e_{t-1} = b_{j_{t-1}}$. Since the considered rule is true, the rule

$$\bigwedge_{i \in \{1, \ldots, m\} \setminus \{j_t\}} (a_i(x) = b_i) \rightarrow a_{j_t}(x) = e_t$$

is true, too. But this is impossible since $u \in I(S)$. Thus, u belongs to $Ext^3(S)$.

Let $u = (b_1, \ldots, b_m) \in V(S) \setminus (U \cup I(S))$. Let us show that u does not belong to $Ext^3(S)$. Since $u \notin U$, the system of equations (1.5) has no solutions from U.

Since $u \notin I(S)$, there exists $j \in \{1, \ldots, m\}$ such that the system of equations (1.6) either has no solutions from U, or there exists $c \in V_{a_j} \setminus \{b_j\}$ such that, for any solution $v = (c_1, \ldots, c_m) \in U$ of this system, $c_j = c$.

In the first case, the rule

$$\bigwedge_{i \in \{1, \ldots, m\} \setminus \{j\}} (a_i(x) = b_i) \rightarrow a_j(x) = e \,, \tag{1.7}$$

where $e \in V_{a_j} \setminus \{b_j\}$, is true for any object from U but not for u.

In the second case, the rule

$$\bigwedge_{i \in \{1, \ldots, m\} \setminus \{j\}} (a_i(x) = b_i) \rightarrow a_j(x) = c \tag{1.8}$$

is true for any object from U but not for u. Thus, u does not belong to $Ext^3(S)$. $\square$

Corollary 1.16. *For each object v from $Ext^3(S)$ there exists an object u from U such that v and u are different on at most one component, i.e., the Hamming distance between any object $v \in Ext^3(S)$ and the set U is at most 1.*

Corollary 1.17. $\left|Ext^3(S)\right| \leq n + nm(k - 1)$, *where* $n = |U|$, $m = |A|$ *and* $k = \max\left\{\left|V_{a_j}\right| : a_j \in A\right\}$.

Proof. Using Theorem 1.15 we obtain $Ext^3(S) = U \cup I(S)$. From Corollary 1.16 it follows that $|I(S)| \leq nm(k - 1)$. Therefore, $\left|Ext^3(S)\right| \leq n + nm(k - 1)$. $\square$

Example 1.18. Let $S = (U, A)$ be the information system such that

$$U = \{(0, 1), (0, 2), (1, 0), (1, 2), (2, 0), (2, 1)\} \,,$$

$A = \{a_1, a_2\}$, and, for any $a_j \in A$ and $u \in \{0, 1, 2\}^2$, the value $a_j(u)$ is equal to the j-th component of u. Then

$$I(S) = \{(0, 0), (1, 1), (2, 2)\} \,.$$

1.4.3 Checking Membership to $Ext^3(S)$ and Construction of $Ext^3(S)$

We now describe a polynomial algorithm $\mathcal{A}_6$ which, for a given element $u \in V(S)$, recognizes if this element belongs to $I(S)$, or not.

Let us consider a polynomial algorithm $\mathcal{A}_7$ for the set $I(S)$ construction.

We now describe a polynomial algorithm $\mathcal{A}_8$ for the set $Ext^3(S)$ construction.

1.4.4 Description of $Ext^3(S)$

Let $F = (a_1, \ldots, a_m)$ and $\Gamma = \Gamma(S, F)$. We now consider a polynomial algorithm $\mathcal{A}_9$ which transforms the set $I(S)$ and the set of equation systems $Eq(\Gamma)$ into a set $R^3(\Gamma)$ of true deterministic rules.

Algorithm 1.6. Algorithm $\mathcal{A}_6$ for recognition of membership to $I(S)$

Input   : Information system $S = (U, A)$, and element $u = (b_1, \ldots, b_m) \in V(S)$.
Output: "Yes" if $u \in I(S)$, and "no" otherwise.
if $u \notin U$ *and for each* $j \in \{1, \ldots, m\}$ *the system of equations (1.6) has two*
solutions $u_1 = (c_1, \ldots, c_m)$ *and* $u_2 = (d_1, \ldots, d_m)$ *from* U *such that* $c_j \neq b_j$,
$d_j \neq b_j$ *and* $c_j \neq d_j$ **then**
| return "yes";
else
| return "no";
end

Algorithm 1.7. Algorithm $\mathcal{A}_7$ for construction $I(S)$

Input   : Information system $S = (U, A)$.
Output: Set $I(S)$.
$I(S) \longleftarrow \emptyset$;
for $(c_1, \ldots, c_m) \in U$; $j \in \{1, \ldots, m\}$; $c \in V_{a_j} \setminus \{c_j\}$ **do**
| $u \longleftarrow (c_1, \ldots, c_{j-1}, c, c_{j+1}, \ldots, c_m)$;
| apply the algorithm $\mathcal{A}_6$ to the information system S and object u;
| **if** *the output of* $\mathcal{A}_6$ *is "yes"* **then**
| | $I(S) \longleftarrow I(S) \cup \{u\}$;
| **end**
end

Algorithm 1.8. Algorithm $\mathcal{A}_8$ for construction $Ext^3(S)$

Input   : Information system $S = (U, A)$.
Output: Set $Ext^3(S)$.
construct the set $I(S)$ by the algorithm $\mathcal{A}_7$;
$Ext^3(S) \longleftarrow I(S) \cup U$;

Lemma 1.19. *The set of rules* $R^3(\Gamma) = R^3(\Gamma(S, F))$ *has the following properties:*

1. $R^3(\Gamma) \subseteq Rul^3(S)$.
2. *The set of objects from* $V(S)$, *for which any rule from* $R^3(\Gamma)$ *it true, coincides with* $Ext^3(S)$.
3. $|R^3(\Gamma)| \leq 2mn(k-1)$, *where* $k = \max\{|V_{a_j}| : a_j \in A\}$.

Proof. From Lemma 1.5 it follows that for each system of equations from $Eq(\Gamma)$ any solution of this system from $V(S)$ belongs to $V(S) \setminus U$, for each object $u \in V(S) \setminus U$ there exists a system from $Eq(\Gamma)$ such that u is a solution of this system, and $|Eq(\Gamma)| \leq mn(k-1)$.

Let $u \in I(S)$. One can show that if u is a solution of an equation system (1.3) from $Eq(\Gamma)$, then in this system $t = m$, and u is the unique solution of this system from $V(S)$. For each $u \in I(S)$ we remove from $Eq(\Gamma)$ the system

---

Algorithm 1.9. Algorithm $\mathcal{A}_9$ for construction $R^3(\Gamma)$

Input   : Set $I(S)$ and set of equation systems $Eq(\Gamma)$, where $\Gamma = \Gamma(S, F)$ and
$\qquad\qquad F = (a_1, \ldots, a_m)$.
Output: Set $R^3(\Gamma)$ of true deterministic rules.
$R^3(\Gamma) \longleftarrow \emptyset$;
$Eq^*(\Gamma) \longleftarrow Eq(\Gamma)$;
for $(d_1, \ldots, d_m) \in I(S)$ **do**
$\quad|\quad$ remove from $Eq^*(\Gamma)$ the system of equations $\{a_1(x) = d_1, \ldots, a_m(x) = d_m\}$;
end
for $\{a_1(x) = b_1, \ldots, a_t(x) = b_t\} \in Eq^*(\Gamma)$ **do**
$\quad$ **if** $t < m$ **then**
$\qquad|\quad$ choose $c_1, c_2 \in V_{a_{t+1}}$ such that $c_1 \neq c_2$;
$\qquad|\quad$ add the following rules to the set $R^3(\Gamma)$:

$$(a_1(x) = b_1) \wedge \ldots \wedge (a_t(x) = b_t) \to a_{t+1}(x) = c_1 , \qquad (1.9)$$

$$(a_1(x) = b_1) \wedge \ldots \wedge (a_t(x) = b_t) \to a_{t+1}(x) = c_2 ; \qquad (1.10)$$

$\quad$ **end**
$\quad$ **if** $t = m$ **then**
$\qquad|\quad$ choose $j \in \{1, \ldots, m\}$ such that either (1.6) has no solutions from U, or
$\qquad|\quad$ there exists $c \in V_{a_j} \setminus \{b_j\}$ such that $c_j = c$ for any solution $(c_1, \ldots, c_m)$ of
$\qquad|\quad$ (1.6) from U;
$\quad$ **end**
$\quad$ **if** (1.6) has no solutions from U **then**
$\qquad|\quad$ choose $e \in V_{a_j} \setminus \{b_j\}$;
$\qquad|\quad$ add the rule (1.7) to $R^3(\Gamma)$;
$\quad$ **else**
$\qquad|\quad$ add the rule (1.8) to $R^3(\Gamma)$;
$\quad$ **end**
end

---

of equations which has u as a solution. We denote the obtained set of equation systems by $Eq^*(\Gamma)$.

It is clear that for each system of equations from $Eq^*(\Gamma)$ any solution of this system from $V(S)$ belongs to $V(S) \setminus Ext^3(S)$, for each object $u \in V(S) \setminus Ext^3(S)$ there exists a system from $Eq^*(\Gamma)$ such that u is a solution of this system, and $|Eq^*(\Gamma)| \leq mn(k - 1)$.

Constructing the set $R^3(\Gamma)$, for each system of equations from $Eq^*(\Gamma)$, we add one or two rules to $R^3(\Gamma)$.

Let us consider an arbitrary system of equations (1.3) from $Eq^*(\Gamma)$.

Let $t < m$, and c_1 and c_2 be two different elements from the set $V_{a_{t+1}}$. Then an object $u \in V(S)$ is a solution of (1.3) if and only if at least one of the rules (1.9) and (1.10) is not true for u. It is clear that the considered rules belong to $Rul^3(S)$. We add these two rules to $R^3(\Gamma)$.

Let $t = m$. Let us denote $u = (b_1, \ldots, b_m)$. It is clear that $u \notin U$ and $u \notin I(S)$. Since $u \notin U$, the system of equations (1.5) has no solutions from U. Since

$u \notin I(S)$, there exists $j \in \{1, \ldots, m\}$ such that either the system of equations (1.6) has no solutions from U, or there exists $c \in V_{a_j} \setminus \{b_j\}$ such that, for any solution $v = (c_1, \ldots, c_m) \in U$ of this system, $c_j = c$.

In the first case, the rule (1.7), where $e \in V_{a_j} \setminus \{b_j\}$, belongs to $Rul^3(S)$ and is not true for u. We add this rule to $R^3(\Gamma)$.

In the second case, the rule (1.8) belongs to $Rul^3(S)$ and is not true for u. We add this rule to $R^3(\Gamma)$.

It is clear now that $\left|R^3(\Gamma)\right| \le 2mn(k-1)$, $R^3(\Gamma) \subseteq Rul^3(S)$, and the set of objects from $V(S)$, for which each rule from $R^3(\Gamma)$ is true, coincides with $Ext^3(S) = U \cup I(S)$.   □

Theorem 1.20. *There exists a polynomial algorithm constructing, for a given information system $S = (U, A)$, a set R of rules with the following properties:*

1. *$R \subseteq Rul^3(S)$.*
2. *The set of all objects from $V(S)$, for which any rule from R it true, coincides with $Ext^3(S)$.*
3. *$|R| \le 2mn(k-1)$, where $m = |A|$, $n = |U|$ and $k = \max\left\{\left|V_{a_j}\right| : a_j \in A\right\}$.*

Proof. The considered algorithm is the following. Let $A = \{a_1, \ldots, a_m\}$. Set $F = (a_1, \ldots, a_m)$. Using algorithm $\mathcal{A}_2$ we construct the decision tree $\Gamma = \Gamma(S, F)$. Next, using algorithms $\mathcal{A}_3$ and $\mathcal{A}_7$ we construct the set of equation systems $Eq(\Gamma)$ and the set $I(S)$. Finally, using algorithm $\mathcal{A}_9$ we construct the set of rules $R^3(\Gamma)$. Set $R = R^3(\Gamma)$.

It is clear that the considered algorithm has polynomial time complexity. From Lemma 1.19 it follows that the system of rules R has properties formulated in the theorem.   □

1.5   True Realizable Deterministic Rules

Let S be an information system. By $Rul^4(S)$ we denote the set of deterministic rules which are true and realizable. By $Ext^4(S)$ we denote the set of objects from $V(S)$, for which each rule from $Rul^4(S)$ is true. The set $Ext^4(S)$ is called *maximal consistent extension of U relative to the set of rules $Rul^4(S)$*.

1.5.1   Information Systems with at Most One Non-binary Attribute

An attribute $a_j \in A$ is called *non-binary* if $\left|V_{a_j}\right| > 2$. In this subsection, we extend a bit the following result from [85]: if an information system $S = (U, A)$ contains at most one non-binary attribute, then $Ext^4(S) = U$. We show that if the set of attributes $C(S)$ contains at most one non-binary attribute, then $Ext^4(S) = U$, and there exists a polynomial algorithm constructing, for such information systems, relatively small systems of true realizable deterministic rules describing exactly the set $Ext^4(S)$.

Let us assume that for the information system S the set $C(S)$ contains at most one non-binary attribute. Let, for the definiteness, $C(S) = \{a_1, \ldots, a_p\}$,

---

Algorithm 1.10. Algorithm $\mathcal{A}_{10}$ for construction $R^4(\Gamma)$

Input : Set of equation systems $Eq(\Gamma)$.
Output: Set $R^4(\Gamma)$ of deterministic rules.
$R^4(\Gamma) \longleftarrow \emptyset$;
for *each equation system* $\{a_1(x) = b_1, \ldots, a_t(x) = b_t\}$ *from* $Eq(\Gamma)$ **do**
$\quad$ choose $b'_t \in V_{a_t} \setminus \{b_t\}$;
$\quad$ add rule $(a_1(x) = b_1) \wedge \ldots \wedge (a_{t-1}(x) = b_{t-1}) \rightarrow a_t(x) = b'_t$ to $R^4(\Gamma)$;
end

---

and a_1 be the non-binary attribute, if $C(S)$ contains a non-binary attribute. Let us denote $F = (a_1, \ldots, a_p)$ and $\Gamma = \Gamma(S, F)$. We now consider a polynomial algorithm $\mathcal{A}_{10}$ which transforms the set of equation systems $Eq(\Gamma)$ into a set $R^4(\Gamma)$ of deterministic rules.

Note that from Lemma 1.5 it follows that $t \geq 2$. Therefore, a_t is a binary attribute.

Lemma 1.21. *The set of rules* $R^4(\Gamma) = R^4(\Gamma(S, F))$ *has the following properties:*

1. $R^4(\Gamma) \subseteq Rul^4(S)$.
2. *Rules from* $R^4(\Gamma)$ *use only attributes from* F.
3. *The set of objects from* $V(S)$, *for which any rule from* $R^4(\Gamma)$ *it true, coincides with* U.
4. $|R^4(\Gamma)| \leq pn(k' - 1) \leq mn(k - 1)$, *where* p *is the number of attributes in* F, $k' = \max\{|V_{a_j}| : a_j \in F\}$ *and* $k = \max\{|V_{a_j}| : a_j \in A\}$.

Proof. From the descriptions of the decision tree Γ and the set $Eq(\Gamma)$ it follows that any equation system (1.3) from $Eq(\Gamma)$ has no solutions from U, but the system obtained from (1.3) by removal of last equation has a solution from U. Therefore, $R^4(\Gamma) \subseteq Rul^4(S)$. Hence, each rule from $R^4(\Gamma)$ is true for each element from U. From Lemma 1.5 it follows that for each element from $V(S) \setminus U$ there exists a rule from $R^4(\Gamma)$ which is not true for this element. Thus, the set of objects from $V(S)$, for which any rule from $R^4(\Gamma)$ it true, coincides with U. The inequalities $|R^4(\Gamma)| \leq pn(k' - 1) \leq mn(k - 1)$ follow from Lemma 1.5. It is clear that rules from $R^4(\Gamma)$ use only attributes from F. $\qquad\square$

Theorem 1.22. *There exists a polynomial algorithm which, for a given information system* $S = (U, A)$ *such that* $C(S)$ *contains at most one non-binary attribute, constructing a set* R *of rules with the following properties:*

1. $R \subseteq Rul^4(S)$.
2. *Rules from* R *use only attributes from* $C(S)$.
3. *The set of all objects from* $V(S)$, *for which any rule from* R *it true, coincides with* U.
4. $|R| \leq pn(k' - 1) \leq mn(k - 1)$, *where* $p = |C(S)|$, $m = |A|$, $n = |U|$, $k' = \max\{|V_{a_j}| : a_j \in C(S)\}$, *and* $k = \max\{|V_{a_j}| : a_j \in A\}$.

Proof. The considered algorithm is the following. Using algorithm $\mathcal{A}_1$ we construct the unique irreducible separating set $C(S)$ which is the core and unique reduct for the decision system $D = (V(S), A, d)$ corresponding to S. Let us assume that $C(S)$ contains at most one non-binary attribute, $C(S) = \{a_{j_1}, \ldots, a_{j_p}\}$ and $a_{j_2}, \ldots, a_{j_p}$ are binary attributes. Set $F = (a_{j_1}, \ldots, a_{j_p})$. Using algorithm $\mathcal{A}_2$ we construct the decision tree $\Gamma = \Gamma(S, F)$. Next, using algorithms $\mathcal{A}_3$ and $\mathcal{A}_{10}$ we construct the set of equation systems $Eq(\Gamma)$ and the set of rules $R^4(\Gamma)$. Set $R = R^4(\Gamma)$.

It is clear that the considered algorithm has polynomial time complexity. From Lemma 1.21 it follows that the system of rules R has properties formulated in the theorem. □

Corollary 1.23. *For any information system $S = (U, A)$, where $C(S)$ contains at most one non-binary attribute, the equality $Ext^4(S) = U$ holds.*

The following example shows that there exists information system $S = (U, A)$, where $C(S)$ contains exactly two non-binary attributes and $Ext^4(S) \neq U$. This information system was considered in [85].

Example 1.24. Let $S = (U, A)$, where $U = \{(0, 1), (1, 0), (0, 2), (2, 0)\}$, $A = \{a_1, a_2\}$, and for any $a_j \in A$ and any $u \in \{0, 1, 2\}^2$ the value $a_j(u)$ is equal to j-th component of u. It is clear that a_1 and a_2 are non-binary attributes. One can show that $C(S) = A$ and $Ext^4(S) \setminus U = \{(0, 0)\}$.

1.5.2   Structure of $Ext^4(S)$

In this subsection, we prove that there exist information systems $S = (U, A)$, for which the cardinality of the set $Ext^4(S) \setminus U$ is huge relative to the cardinality of U, and there exist information systems $S = (U, A)$ with large Hamming distance between $Ext^4(S) \setminus U$ and U.

Proposition 1.25. *For any natural $m \geq 3$, there exists an information system $S = (U, A)$ such that $|A| = m$, $|U| = 2m + 2$, $V(S) = \{0, 1, 2\}^m$ and $\left|Ext^4(S)\right| = 2^m + 2m + 1$.*

Proof. Let $m \geq 3$, $U \subseteq \{0, 1, 2\}^m$,

$$U = \{(0, 0, \ldots, 0), (1, 0, \ldots, 0), \ldots, (0, 0, \ldots, 1),$$
$$(2, 2, \ldots, 2), (1, 2, \ldots, 2), \ldots, (2, 2, \ldots, 1)\},$$

$A = \{a_1, \ldots, a_m\}$, and for any $a_j \in A$ and any $u \in \{0, 1, 2\}^m$ the value $a_j(u)$ be equal to the j-th component of u. It is clear that $|A| = m$, $|U| = 2m + 2$ and $V(S) = \{0, 1, 2\}^m$.

Let i, j, l be pairwise different numbers from $\{1, \ldots, m\}$. One can show that the set of rules $Rul^4(S)$ contains the following rules:

$$(a_i(x) = 0) \wedge (a_j(x) = 1) \rightarrow a_l(x) = 0 , \tag{1.11}$$

$$(a_i(x) = 2) \wedge (a_j(x) = 1) \rightarrow a_l(x) = 2 . \tag{1.12}$$

Let $J = Ext^4(S) \setminus U$ and

$$K = \{(1,\ldots,1)\} \cup (\{0,2\}^m \setminus \{(0,\ldots,0),(2,\ldots,2)\}) .$$

Since (1.11), (1.12) belong to $Rul^4(S)$ for any pairwise different $i,j,l \in \{1,\ldots,m\}$, one can show that $J \subseteq K$.

It is not difficult to prove that the set $Rul^4(S)$ does not contain rule (1.2), if at least one of the following conditions holds: (i) $t = 1$; (ii) $b_1,\ldots,b_{t-1} \in \{0,2\}$; (iii) $b_1 = \ldots = b_{t-1} = 1$.

Using these facts one can show that $K \subseteq J$. Therefore, $K = J$, $Ext^4(S) = U \cup K$ and $\left| Ext^4(S) \right| = 2^m + 2m + 1$. $\qquad \square$

Proposition 1.26. *For any natural $m \geq 3$, there exists an information system $S = (U, A)$ such that $|A| = m$, $|U| = 3m + 2$, $\left| Ext^4(S) \right| = 3m + 3$, and the Hamming distance between sets U and $Ext^4(U) \setminus U$ is equal to $m - 1$.*

Proof. Let $m \geq 3$, $U \subseteq \{0,1,2\}^m$,

$$\begin{aligned} U = \{&(0,0,\ldots,0),(1,0,\ldots,0),\ldots,(0,0,\ldots,1), \\ &(2,2,\ldots,2),(1,2,\ldots,2),\ldots,(2,2,\ldots,1), \\ &(2,0,\ldots,0),\ldots,(0,0,\ldots,2)\} , \end{aligned}$$

$A = \{a_1,\ldots,a_m\}$, and for any $a_j \in A$ and any $u \in \{0,1,2\}^m$ the value $a_j(u)$ be equal to the j-th component of u. It is clear that $|A| = m$, $|U| = 3m + 2$ and $V(S) = \{0,1,2\}^m$.

Let i,j,l be pairwise different numbers from $\{1,\ldots,m\}$. One can show that the set of rules $Rul^4(S)$ contains rules (1.11), (1.12) and the rule

$$(a_i(x) = 0) \wedge (a_j(x) = 2) \rightarrow a_l(x) = 0 . \tag{1.13}$$

Let $J = Ext^4(S) \setminus U$ and $K = \{(1,\ldots,1)\}$. Since (1.11)–(1.13) belong to $Rul^4(S)$ for any pairwise different $i,j,l \in \{1,\ldots,m\}$, one can show that $J \subseteq K$.

It is not difficult to prove that the set $Rul^4(S)$ does not contain rule (1.2), if at least one of the following conditions holds: (i) $t = 1$; (ii) $b_1 = \ldots = b_{t-1} = 1$.

Using these facts one can show that $K \subseteq J$. Therefore, $K = J$, $Ext^4(S) = U \cup K$ and $\left| Ext^4(S) \right| = 3m + 3$. It is clear that the Hamming distance between sets U and $\{(1,\ldots,1)\}$ is equal to $m - 1$. $\qquad \square$

1.5.3 Algorithmic Problems Related to $Ext^4(S)$

Polynomial algorithms, which recognize if the equality $Ext^4(S) = U$ holds or not for a given information system $S = (U, A)$, are unknown. From Proposition 1.26 it follows that it is useless to look for such algorithms among algorithms based on local search in neighborhoods of objects from U.

The next statement follows immediately from Proposition 1.25:

Theorem 1.27. *There is no polynomial algorithm constructing, for a given information system S, the set $Ext^4(S)$.*

Totally polynomial algorithms for constructing the set $Ext^4(S)$ are unknown (totally polynomial means polynomial in time, depending on the length of input and output). Polynomial algorithms for constructing systems of rules from $Rul^4(S)$ which describe exactly the set $Ext^4(S)$ are also unknown. There is only one exception:

Theorem 1.28. ([70]) *There exists a polynomial algorithm recognizing, for a given information system $S = (U, A)$, and given object u from $V(S) \setminus U$, if this object belongs to $Ext^4(S)$, or not.*

The algorithm considered in [70] is based on a relatively complicated criterion of membership to maximal consistent extension of information system. In [49] a new algorithm was introduced. This algorithm is based on a simpler criterion than the algorithm described in [70]. The criterion under consideration is convenient for theoretical analysis of maximal consistent extensions of information systems relative to true and realizable deterministic rules.

We now describe the polynomial algorithm $\mathcal{A}_{11}$ from [49].

Note that $|P_i^j| \geq 1$ for any $i \in \{1, \ldots, n\}$ and $j \in \{1, \ldots, m\} \setminus M_i$.

The considered algorithm is based on the following criterion.

Algorithm 1.11. Algorithm $\mathcal{A}_{11}$ for recognition of membership to $Ext^4(S)$

Input : Information system $S = (U, A)$, where $U = \{u_1, \ldots, u_n\}$ and
           $A = \{a_1, \ldots, a_m\}$, and object $u \in V(S) \setminus U$.
Output: "Yes" if $u \in Ext^4(S)$, and "no" otherwise.
for $i = 1, \ldots, n$ **do**
| construct the set $M_i = \{j : j \in \{1, \ldots, m\}, a_j(u) = a_j(u_i)\}$;
end
for $i = 1, \ldots, n$; $j \in \{1, \ldots, m\} \setminus M_i$ **do**
| construct the set $P_i^j = \{a_j(u_t) : u_t \in U, a_l(u_t) = a_l(u)$ for each $l \in M_i\}$;
end
if $|P_i^j| \geq 2$ *for any* $i \in \{1, \ldots, n\}$ *and* $j \in \{1, \ldots, m\} \setminus M_i$ **then**
| return "yes";
else
| return "no";
end

Proposition 1.29. ([49]) *The relation $u \notin Ext^4(S)$ holds if and only if $|P_i^j| = 1$ for some $i \in \{1, \ldots, n\}$ and $j \in \{1, \ldots, m\} \setminus M_i$.*

Proof. Let $u = (b_1, \ldots, b_m)$. Let us assume that for some $i \in \{1, \ldots, n\}$ and $j \in \{1, \ldots, m\} \setminus M_i$ the set P_i^j has only one element c. One can show that the rule

$$\bigwedge_{l \in M_i} (a_l(x) = b_l) \to a_j(x) = c$$

belongs to the set $Rul^4(S)$. It is clear that $a_j(u_i) = c$. Since $j \notin M_i$, we have $b_j \neq c$. Therefore, the considered rule is not true for the element u. Thus, $u \notin Ext^4(S)$.

Let $u \notin Ext^4(S)$. Then there exists a rule

$$(a_{l_1}(x) = c_1) \wedge \ldots \wedge (a_{l_s}(x) = c_s) \to a_j(x) = c$$

from $Rul^4(S)$ which is not true for u. We denote this rule by r. Since r is not true for u, we have $a_{l_1}(u) = c_1, \ldots, a_{l_s}(u) = c_s, a_j(u) \neq c$. Since $r \in Rul^4(S)$, there exists an element u_i from U such that $a_{l_1}(u_i) = c_1, \ldots, a_{l_s}(u_i) = c_s, a_j(u_i) = c$. It is clear that $l_1, \ldots, l_s \in M_i$ and $j \notin M_i$. Let $P = \{a_j(u_t) : u_t \in U, a_{l_1}(u_t) = c_1, \ldots, a_{l_s}(u_t) = c_s\}$. Since $r \in Rul^4(S)$, we have $P = \{c\}$. Taking into account that $l_1, \ldots, l_s \in M_i$ we conclude that $P_i^j \subseteq P$. Thus, $|P_i^j| = 1$. $\square$

1.6 Conclusions

We study maximal consistent extensions of information systems relative to the set of true realizable inhibitory rules, the set of true inhibitory rules, the set of true deterministic rules, and the set of true realizable deterministic rules. We consider the following algorithmic problems: (i) the membership to the extension, (ii) the construction of the extension, and (iii) the construction of rule system describing the extension. For the first three kinds of extensions, the considered algorithmic problems can be solved effectively. For the fourth kind of extensions we know an effective polynomial algorithm only for the problem (i). For the problem (ii) there are no polynomial algorithms. For the problem (iii) polynomial algorithms are unknown.

The obtained results show that the inhibitory rules provide an essentially more powerful tool for knowledge representation than the deterministic rules. These results will be useful in applications of information systems for analysis and design of concurrent systems specified by data tables, as well as for problems of classification and prediction.

2 Minimal Inhibitory Association Rules for Almost All k-Valued Information Systems

There are three approaches to use inhibitory rules in classifiers: (i) lazy algorithms based on an information about the set of all inhibitory rules, (ii) standard classifiers based on a subset of inhibitory rules constructed by a heuristic, and (iii) standard classifiers based on the set of all minimal (irreducible) inhibitory rules. The aim of this chapter is to show that the last approach is not feasible (from computational complexity point of view).

We restrict our considerations to the class of k-valued information systems, i.e., information systems with attributes having values from $\{0, \ldots, k-1\}$, where $k \geq 2$. Note that the case $k = 2$ was considered earlier in [51].

We show that, for any positive real number α, for almost all k-valued information systems S with m attributes and $n = \lfloor m^\alpha \rfloor$ objects the length of each minimal (irreducible) inhibitory association rule is relatively short. This means that the length is logarithmic relative to the number m. We also prove that for almost all k-valued information systems with m attributes and $n = \lfloor m^\alpha \rfloor$ objects the number of minimal inhibitory association rules is at least $m^{(1/5)\alpha \log_k m}$ and at most $m^{2(1+\alpha)\log_k m+7}$. Hence, from this result it follows that for almost all k-valued information systems with the polynomial number of objects in the number of attributes the number of minimal inhibitory rules is not polynomial in the number of attributes.

Based on the obtained bounds we prove for k-valued information systems with m attributes and $n = \lfloor m^\alpha \rfloor$ objects that

- there is no algorithm which for almost all k-valued information systems constructs all minimal inhibitory association rules and for such information systems has polynomial time complexity relative to the length of input,
- there exists an algorithm which for almost all k-valued information systems constructs all minimal inhibitory association rules and for such information systems has polynomial time complexity relative to the length of input and the length of output.

The obtained results have interesting interpretation for k-valued decision systems (decision tables). If for any considered information system S one attribute

P. Delimata et al.: Inhibitory Rules in Data Analysis, SCI 163, pp. 31–42.
springerlink.com   © Springer-Verlag Berlin Heidelberg 2009

is selected as a decision attribute we obtain a k-valued decision system T [60]. Minimal inhibitory rules of S with the selected decision attribute on their right-hand sides are called the minimal inhibitory decision rules of T. Such rules can be used for construction of classifiers. Let us consider k-valued information systems S with m attributes and $n = \lfloor m^\alpha \rfloor$ objects. Then, for almost all such systems S all minimal inhibitory decision rules for each decision system obtained from S are relatively short (i.e., the length of each rule is logarithmic relative to the number m). However, for at least one decision system obtained from S the number of minimal inhibitory decision rules is at least $m^{(1/5)\alpha \log_k m - 1}$ (i.e., is not polynomial in the number of conditional attributes and the number of objects).

From these facts it follows that either it is necessary to use some heuristics for generating from a given information or decision system only a subset of "important" rules, or it is necessary to use lazy classification algorithms.

The study of minimal inhibitory association rules over S is based on investigation of the so-called irreducible inconsistent systems of equations (descriptors) over S. The results for irreducible inconsistent systems of equations can be also used for proving some other properties.

This chapter is based on papers [51, 53]. Note that some similar results were obtained in [36, 37, 38, 46] for partial (not for exact) covers, reducts and decision rules.

The chapter contains seven subsections. In Sect. 2.1, basic concepts are introduced. In Sect. 2.2, relationships between minimal inhibitory association rules and irreducible inconsistent systems of equations are presented. Sections 2.3–2.5 include results on irreducible inconsistent systems of equations and Sect. 2.6 – on minimal inhibitory association rules. Short conclusions are included in Sect. 2.7.

2.1 Main Notions

Let k, m and n be natural numbers and $k \geq 2$. A *k-valued information system with m attributes and n objects* is a table S with m columns labeled with symbols $a_1, \ldots, a_m$ and n rows labeled with symbols $u_1, \ldots, u_n$. This table is filled by numbers from $\{0, \ldots, k-1\}$. Symbols $a_1, \ldots, a_m$ are interpreted as names of attributes, and symbols $u_1, \ldots, u_n$ – as names of objects. For $i = 1, \ldots, n$ and $j = 1, \ldots, m$ the value at the intersection of the row u_i and the column a_j is interpreted as the value $a_j(u_i)$ (the value of the attribute a_j on the object u_i). The information system S can be represented as a pair (U, A), where $U = \{u_1, \ldots, u_n\}$ is the set of objects, $A = \{a_1, \ldots, a_m\}$ is the set of attributes, and the values of attributes from A on objects from U are given by the considered table. In this case, S is called the *tabular representation* of (U, A). By $I_k(m, n)$ we denote the set of all k-valued information systems with m attributes and n objects. The number of information systems in $I_k(m, n)$ is equal to k^{mn}.

A rule

$$(a_{j_1}(x) = b_1) \wedge \ldots \wedge (a_{j_{t-1}}(x) = b_{t-1}) \rightarrow a_{j_t}(x) \neq b_t \, ,$$

where $t \geq 1$, $a_{j_1}, \ldots, a_{j_t} \in A$, $b_1, \ldots, b_t \in \{0, \ldots, k-1\}$, and numbers $j_1, \ldots, j_t$ are pairwise different, is called a *minimal inhibitory association* rule over S if

(i) this rule is true for all objects from U; (ii) if $t \geq 2$, then there exists an object from U for which the left-hand side of the rule is true; (iii) if $t \geq 2$ and we remove an arbitrary condition (equation) from the left-hand side of the considered rule we obtain a rule which is not true for at least one object from U. The number t is called the *length* of the considered rule. Later we will omit the word "association".

A system of equations

$$\{a_{j_1}(x) = b_1, \ldots, a_{j_t}(x) = b_t\} ,$$

where $t \geq 1$, $a_{j_1}, \ldots, a_{j_t} \in A$, $b_1, \ldots, b_t \in \{0, \ldots, k-1\}$ and numbers $j_1, \ldots, j_t$ are pairwise different, is called an *irreducible inconsistent* system of equations over S if (i) the considered system is *inconsistent* – has no solutions from U; (ii) each proper subsystem of this system is *consistent* – has a solution from U (by definition, empty set of equations is consistent). The number t is called the *cardinality* of the considered equation system.

Let $\mathcal{P}$ be a property of information systems and let $P_k(m, n)$ be the number of information systems from $I_k(m, n)$ for which $\mathcal{P}$ holds. The number

$$\frac{P_k(m, n)}{k^{mn}}$$

is called the *fraction* of information systems from $I_k(m, n)$ for which the property $\mathcal{P}$ holds.

Let α be a positive real number. We also consider k-valued information systems with m attributes and $n = \lfloor m^\alpha \rfloor$ objects, i.e., information systems from the set $I_k(m, \lfloor m^\alpha \rfloor)$. We say that the property $\mathcal{P}$ holds *for almost all* information systems from $I_k(m, \lfloor m^\alpha \rfloor)$ if the fraction

$$\frac{P_k(m, \lfloor m^\alpha \rfloor)}{k^{m \lfloor m^\alpha \rfloor}}$$

of information systems from $I_k(m, \lfloor m^\alpha \rfloor)$, for which the property $\mathcal{P}$ holds, tends to 1 as m tends to infinity.

Later we will assume that some natural k, $k \geq 2$, is fixed.

2.2 Relationships between Rules and Inconsistent Equation Systems

There exists simple correspondence between minimal inhibitory rules and irreducible inconsistent systems of equations. We omit the proof of the following statement.

Proposition 2.1. *Let* $S = (U, A) \in I_k(m, n)$, $t \geq 1$, $a_{j_1}, \ldots, a_{j_t} \in A$, $b_1, \ldots, b_t \in \{0, \ldots, k-1\}$, *numbers* $j_1, \ldots, j_t$ *be pairwise different and* $l \in \{1, \ldots, t\}$. *Then the system of equations*

$$\{a_{j_1}(x) = b_1, \ldots, a_{j_t}(x) = b_t\}$$

is an irreducible inconsistent system of equations over S if and only if the rule

$$\bigwedge_{i \in \{1,\dots,t\} \setminus \{l\}} (a_{j_i}(x) = b_i) \rightarrow a_{j_l}(x) \neq b_l$$

is a minimal inhibitory rule over S.

Let us denote by $M(S)$ the number of minimal inhibitory rules over S, and by $I(S)$ – the number of irreducible inconsistent equation systems over S.

Corollary 2.2. *Let $S \in I_k(m,n)$. Then*

1. *The minimal cardinality of an irreducible inconsistent equation system over S is equal to the minimal length of a minimal inhibitory rule over S.*
2. *The maximal cardinality of an irreducible inconsistent equation system over S is equal to the maximal length of a minimal inhibitory rule over S.*
3. *$I(S) \leq M(S) \leq mI(S)$.*

2.3 Cardinality of Irreducible Inconsistent Equation Systems

In this section, we consider lower and upper bounds on cardinality of irreducible inconsistent equation systems for k-valued information systems with n objects and m attributes. Under some assumptions on m and n, we evaluate the fraction of information systems for which the considered bounds hold for any irreducible inconsistent equation system. Also, under some other assumptions on m and n, we prove the existence of large number of inconsistent equation systems of some cardinality.

Theorem 2.3. *Let m, n, q be natural numbers and let*

$$\tau_q(m,n) = \lceil \log_k m + \log_k n + 2 + q \rceil .$$

Then the fraction of information systems from $I_k(m,n)$ having irreducible inconsistent equation systems with cardinality at least $\tau_q(m,n)$ is at most

$$\frac{1}{k^{q(\log_k m + \log_k n + 2 + q)}} .$$

Proof. Let $\tau = \tau_q(m,n)$ and let us assume that $S = (U,A) \in I_k(m,n)$ has an irreducible inconsistent equation system

$$\{a_{j_1}(x) = b_1, \dots, a_{j_p}(x) = b_p\} ,$$

where $p \geq \tau$, $a_{j_1}, \dots, a_{j_p} \in A$, $b_1, \dots, b_p \in \{0,1\}$ and the numbers $j_1, \dots, j_p$ are pairwise different. Then in the tabular representation of S there exist p rows $u_{i_1}, \dots, u_{i_p}$ which at the intersection with columns $a_{j_1}, \dots, a_{j_p}$ have values

$$c_1, b_2, \dots, b_p$$
$$b_1, c_2, \dots, b_p$$
$$\dots$$
$$b_1, b_2, \dots, c_p ,$$

where $c_1 \neq b_1, \ldots, c_p \neq b_p$. Since $p \geq \tau$, in the tabular representation of S there exist τ rows $u_{i_1}, \ldots, u_{i_\tau}$ which at the intersection with columns $a_{j_1}, \ldots, a_{j_\tau}$ have values

$$c_1, b_2, \ldots, b_\tau$$
$$b_1, c_2, \ldots, b_\tau \qquad (2.1)$$
$$\ldots$$
$$b_1, b_2, \ldots, c_\tau \ .$$

It is clear that the number of information systems for which in the tabular representation at the intersection of rows $u_{i_1}, \ldots, u_{i_\tau}$ with columns $a_{j_1}, \ldots, a_{j_\tau}$ there are values (2.1) is at most $k^{mn-\tau^2}$. There are k^τ variants for the choice of numbers $b_1, \ldots, b_\tau$, $(k-1)^\tau$ variants for the choice of numbers $c_1, \ldots, c_\tau$, at most m^τ variants for the choice of attributes $a_{j_1}, \ldots, a_{j_\tau}$, and at most n^τ variants for the choice of objects $u_{i_1}, \ldots, u_{i_\tau}$. Therefore, the number of information systems from $I_k(m,n)$ having irreducible inconsistent equation systems with cardinality at least τ is at most

$$k^{mn+2\tau+\tau\log_k m+\tau\log_k n-\tau^2} = k^{mn+\tau(2+\log_k m+\log_k n-\tau)} \leq k^{mn-q\tau} \ .$$

Therefore, the fraction of the considered information systems is at most

$$\frac{k^{mn-q\tau}}{k^{mn}} = \frac{1}{k^{q\tau}} \leq \frac{1}{k^{q(\log_k m+\log_k n+2+q)}} \ . \qquad \square$$

Note that with linear increase of q we have exponential decrease of the fraction of information systems that have irreducible inconsistent equation systems which cardinality is at least $\tau_q(m,n)$.

Theorem 2.4. *For $m \geq k$, $n \geq k$, and a natural number q we define the following function:*

$$\varrho_q(m,n) = \lfloor \log_k n - \log_k \log_k n - \log_k \log_k m - 1 - q \rfloor \ .$$

If (m,n) is satisfying the condition $1 \leq \varrho_q(m,n) \leq m$, then the fraction of information systems from $I_k(m,n)$ having inconsistent equation systems with cardinality at most $\varrho_q(m,n)$ is at most

$$\frac{1}{k^{(k^{q+1}/\ln k-2)\log_k m \log_k n}} \ .$$

Proof. Let $\varrho = \varrho_q(m,n)$. Let us assume that $S = (U,A) \in I_k(m,n)$ has an inconsistent equation system with at most ϱ equations. In this case, S has an inconsistent equation system with ϱ equations

$$\{a_{j_1}(x) = b_1, \ldots, a_{j_\varrho}(x) = b_\varrho\} \ ,$$

where $a_{j_1}, \ldots, a_{j_\varrho} \in A$, $b_1, \ldots, b_\varrho \in \{0, \ldots, k-1\}$ and numbers $j_1, \ldots, j_\varrho$ are pairwise different. Then in the tabular representation of S each row at the intersection with columns $a_{j_1}, \ldots, a_{j_\varrho}$ has a tuple of values which is different from $(b_1, \ldots, b_\varrho)$.

The number of information systems which in the tabular representation at the intersection of each row with columns $a_{j_1}, \ldots, a_{j_\varrho}$ have tuples of values different from $(b_1, \ldots, b_\varrho)$ is equal to

$$k^{mn-\varrho n}(k^\varrho - 1)^n = k^{mn}\left(\frac{k^\varrho - 1}{k^\varrho}\right)^n = k^{mn}\left(\frac{k^\varrho - 1}{k^\varrho}\right)^{k^\varrho n/k^\varrho}.$$

Using well known inequality

$$\left(\frac{c-1}{c}\right)^c \leq \frac{1}{e},$$

which holds for any natural number c, we obtain

$$k^{mn}\left(\frac{k^\varrho - 1}{k^\varrho}\right)^{k^\varrho n/k^\varrho} \leq k^{mn-n/(k^\varrho \ln k)}.$$

There are exactly k^ϱ variants for the choice of numbers $b_1, \ldots, b_\varrho$, and at most m^ϱ variants for the choice of ϱ columns. Therefore, the number of information systems that have inconsistent equation systems which cardinality is at most ϱ is at most $k^{mn+\varrho+\varrho \log_k m - n/(k^\varrho \ln k)}$. It is clear that

$$\frac{n}{k^\varrho \ln k} \geq \frac{nk^{q+1}\log_k m \log_k n}{n \ln k} = \frac{k^{q+1}\log_k m \log_k n}{\ln k},$$

$\varrho \leq \log_k n$ and $\varrho + \varrho \log_k m \leq \log_k n + \log_k n \log_k m$. Hence,

$$k^{mn+\varrho+\varrho \log_k m - n/(k^\varrho \ln k)} \leq k^{mn-\left(k^{q+1}/\ln k - 2\right)\log_k m \log_k n}.$$

Therefore, the fraction of the considered information systems is at most

$$\frac{k^{mn-\left(k^{q+1}/\ln k - 2\right)\log_k m \log_k n}}{k^{mn}} = \frac{1}{k^{\left(k^{q+1}/\ln k - 2\right)\log_k m \log_k n}}. \qquad \square$$

Note that with linear increase of q we have superexponential decrease of the fraction of information systems that have inconsistent equation systems which cardinality is at most $\varrho_q(m, n)$.

Theorem 2.5. *Let $S = (U, A) \in I_k(m, n)$, $\pi(n) = \lceil \log_k n + 1 \rceil$ and $\pi(n) \leq m$. If $a_{j_1}, \ldots, a_{j_{\pi(n)}} \in A$ are such that the numbers $j_1, \ldots, j_{\pi(n)}$ are pairwise different, then there exist numbers $b_1, \ldots, b_{\pi(n)} \in \{0, \ldots, k-1\}$ for which the system of equations*

$$\{a_{j_1}(x) = b_1, \ldots, a_{j_{\pi(n)}}(x) = b_{\pi(n)}\}$$

is inconsistent.

Proof. Assume the contrary. Then, there exist attributes $a_{j_1}, \ldots, a_{j_{\pi(n)}} \in A$ such that the numbers $j_1, \ldots, j_{\pi(n)}$ are pairwise different and for any numbers $b_1, \ldots, b_{\pi(n)} \in \{0, \ldots, k-1\}$ the system of equations

$$\{a_{j_1}(x) = b_1, \ldots, a_{j_{\pi(n)}}(x) = b_{\pi(n)}\}$$

is consistent. Then $|U| > n$ which is impossible. $\qquad \square$

2.4 Number of Irreducible Inconsistent Equation Systems

In this section, we consider k-valued information systems with m attributes and $n = \lfloor m^\alpha \rfloor$ objects, where α is a positive real number. We present an upper bound on the cardinality of irreducible inconsistent systems of equations, and some lower and upper bounds on the number of irreducible inconsistent systems of equations which hold for almost all information systems from $I_k(m, \lfloor m^\alpha \rfloor)$.

Theorem 2.6. *Let α be a positive real number. Then for almost all information systems $S \in I_k(m, \lfloor m^\alpha \rfloor)$ the cardinality of each irreducible inconsistent equation system is at most $\lceil \log_k m + \log_k \lfloor m^\alpha \rfloor + 3 \rceil - 1$, and*

$$m^{(1/5)\alpha \log_k m} \leq I(S) \leq m^{2(1+\alpha)\log_k m+6} .$$

Proof. Let us assume $\varrho = \lfloor \log_k \lfloor m^\alpha \rfloor - \log_k \log_k \lfloor m^\alpha \rfloor - \log_k \log_k m - 2 \rfloor$, $\tau = \lceil \log_k m + \log_k \lfloor m^\alpha \rfloor + 3 \rceil - 1$, and $\pi = \lceil \log_k \lfloor m^\alpha \rfloor + 1 \rceil$.

Using Theorems 2.3, 2.4 and 2.5 we conclude that for sufficiently large m the fraction of information systems $S = (U, A)$ from $I_k(m, \lfloor m^\alpha \rfloor)$ for which

(a) cardinality of each irreducible inconsistent system of equations is at most τ,
(b) cardinality of each inconsistent system of equations is more than ϱ,
(c) for any attributes $a_{j_1}, \ldots, a_{j_\pi} \in A$ such that the numbers $j_1, \ldots, j_\pi$ are pairwise different there exist numbers $b_1, \ldots, b_\pi \in \{0, \ldots, k-1\}$ for which the system of equations $\{a_{j_1}(x) = b_1, \ldots, a_{j_\pi}(x) = b_\pi\}$ is inconsistent

is at least

$$1 - \frac{1}{k^{\log_k m + \log_k \lfloor m^\alpha \rfloor + 3}} - \frac{1}{k^{(k^2/\ln k - 2)\log_k m \log_k \lfloor m^\alpha \rfloor}} .$$

One can show that $k^2/\ln k - 2 \geq 1$ for any $k \geq 2$. It is clear that for sufficiently large m,

$$1 - \frac{1}{k^{\log_k m + \log_k \lfloor m^\alpha \rfloor + 3}} - \frac{1}{k^{(k^2/\ln k - 2)\log_k m \log_k \lfloor m^\alpha \rfloor}} \geq 1 - \frac{1}{k^{(1+\alpha)\log_k m}} . \quad (2.2)$$

Let us consider an arbitrary information system $S = (U, A) \in I_k(m, \lfloor m^\alpha \rfloor)$ satisfying the conditions (a), (b) and (c).

We now show that for sufficiently large m,

$$m^{(1/5)\alpha \log_k m} \leq I(S) \leq m^{2(1+\alpha)\log_k m+6} . \quad (2.3)$$

Since S satisfies (a), we have for sufficiently large m,

$$I(S) \leq m^\tau k^\tau \leq m^{2\tau} \leq m^{2(1+\alpha)\log_k m+6} .$$

It is clear that each inconsistent system of equations has irreducible inconsistent system as a subsystem (we consider empty set of equations as a consistent system). Let Q be an irreducible inconsistent system of equations. Let us evaluate the number of inconsistent systems of equations of cardinality π which have Q as a subsystem. Let cardinality of Q be equal to t. One can show that $\varrho + 1 \le t \le \tau$. If $t > \pi$, then the considered number is equal to 0. Let $t \le \pi$. There are $C_{m-t}^{\pi-t} k^{\pi-t}$ ways to obtain an inconsistent system of cardinality π from Q by adding equations of the kind $a_j(x) = b$, where a_j does not belong to Q and $b \in \{0, \dots, k-1\}$. It is clear that $C_{m-t}^{\pi-t} k^{\pi-t} \le C_m^{\pi-t} k^{\pi-t}$. If $\pi < m/2$, then $C_m^{\pi-t} k^{\pi-t} \le C_m^{\pi-\varrho} k^{\pi-\varrho}$. Thus, for sufficiently large m the number of inconsistent systems of equations of cardinality π which have Q as a subsystem is at most $C_m^{\pi-\varrho} k^{\pi-\varrho}$.

The number of different inconsistent systems of cardinality π is at least C_m^π. Hence,

$$I(S) \ge \frac{C_m^\pi}{C_m^{\pi-\varrho} k^{\pi-\varrho}} = \frac{(m - \pi + 1) \dots (m - \pi + \varrho)}{k^{\pi-\varrho}(\pi - \varrho + 1) \dots \pi} \ge \frac{1}{k^{\pi-\varrho}} \left(\frac{m - \pi}{\pi} \right)^\varrho .$$

For sufficiently large m,

$$\frac{m - \pi}{\pi} = \frac{m - \lceil \log_k \lfloor m^\alpha \rfloor + 1 \rceil}{\lceil \log_k \lfloor m^\alpha \rfloor + 1 \rceil} \ge m^{1/2} .$$

Therefore,

$$I(S) \ge \frac{m^{\varrho/2}}{k^{\pi-\varrho}} .$$

It is clear that for sufficiently large m the inequality $\varrho \ge (1/2)\alpha \log_k m$ holds. One can show that

$$\pi - \varrho \le \log_k \log_k \lfloor m^\alpha \rfloor + \log_k \log_k m + 5 .$$

Therefore, for sufficiently large m,

$$I(S) \ge \frac{m^{\varrho/2}}{k^{\pi-\varrho}} \ge \frac{m^{(1/4)\alpha \log_k m}}{k^5 \log_k \lfloor m^\alpha \rfloor \log_k m} \ge m^{(1/4)\alpha \log_k m - 3} \ge m^{(1/5)\alpha \log_k m} .$$

Thus, (2.3) holds. Since

$$1 - \frac{1}{k^{(1+\alpha) \log_k m}}$$

(see (2.2)) tends to 1 as m tends to infinity, we conclude that the statement of the theorem holds. $\square$

2.5   Construction of All Irreducible Inconsistent Equation Systems

Let α be a positive real number. We consider k-valued information systems with m attributes and $n = \lfloor m^\alpha \rfloor$ objects. For a given information system

$S = (U, A) \in I_k(m, \lfloor m^\alpha \rfloor)$ it is required to construct all irreducible inconsistent equation systems over S. The length of input for this problem is equal to $m \lfloor m^\alpha \rfloor \lceil \log_2 k \rceil \leq m^{1+\alpha} \lceil \log_2 k \rceil$. The length of output is at least $I(S)$.

Let $\tau(\alpha, m) = \lceil \log_k m + \log_k \lfloor m^\alpha \rfloor + 3 \rceil - 1$. From Theorem 2.6 it follows that for almost all information systems $S \in I_k(m, \lfloor m^\alpha \rfloor)$ the cardinality of each irreducible inconsistent system of equations is at most $\tau(\alpha, m)$, and

$$m^{(1/5)\alpha \log_k m} \leq I(S) \leq m^{2(1+\alpha) \log_k m + 6} .$$

Thus, there is no algorithm which for almost all information systems from $I_k(m, \lfloor m^\alpha \rfloor)$ constructs the set of all irreducible inconsistent systems of equations and for such information systems has polynomial time complexity depending on the length of input.

Let us consider an algorithm which finds all equation systems with cardinality at most $\tau = \tau(\alpha, m)$ and for each such system recognizes if this system is irreducible inconsistent or not. It is clear that this recognition problem can be solved (for one system of equations) in polynomial time depending on the length of input. The number of such systems is at most $k^\tau m^\tau$. One can show that $k^\tau m^\tau \leq m^{3(1+\alpha) \log_k m}$ for sufficiently large m. From Theorem 2.6 it follows that for almost all information systems from $I_k(m, \lfloor m^\alpha \rfloor)$ the considered algorithm finds all irreducible inconsistent equation systems.

The considered algorithm works with at most $m^{3(1+\alpha) \log_k m}$ equation systems for sufficiently large m. Using Theorem 2.6 we conclude that for almost all information systems S from $I_k(m, \lfloor m^\alpha \rfloor)$,

$$m^{3(1+\alpha) \log_k m} = \left(m^{(1/5)\alpha \log_k m} \right)^{15(1+\alpha)/\alpha} \leq (I(S))^{15(1+\alpha)/\alpha} .$$

Thus, there exists an algorithm which for almost all information systems from $I_k(m, \lfloor m^\alpha \rfloor)$ constructs the set of irreducible inconsistent equation systems and for such information systems has polynomial time complexity depending on the length of input and the length of output.

2.6 Minimal Inhibitory Rules

In this section, we translate the results obtained for irreducible inconsistent equation systems to the case of minimal inhibitory rules.

2.6.1 Length of Minimal Inhibitory Rules

In this subsection, we consider some lower and upper bounds on the length of minimal inhibitory rules for k-valued information systems with n objects and m attributes. Under some assumptions on m and n, we evaluate the fraction of information systems for which the considered bounds hold for any minimal inhibitory rule.

The next two statements follow immediately from Theorems 2.3 and 2.4, and from Corollary 2.2.

Theorem 2.7. *Let* m, n, q *be natural numbers and let*

$$\tau_q(m, n) = \lceil \log_k m + \log_k n + 2 + q \rceil .$$

Then the fraction of information systems from $I_k(m, n)$ *having minimal inhibitory rules with the length at least* $\tau_q(m, n)$ *is at most*

$$\frac{1}{k^{q(\log_k m + \log_k n + 2 + q)}} .$$

Theorem 2.8. *For* $m \geq k$, $n \geq k$, *and a natural number* q *we assume,*

$$\varrho_q(m, n) = \lfloor \log_k n - \log_k \log_k n - \log_k \log_k m - 1 - q \rfloor .$$

If (m, n) *satisfies* $1 \leq \varrho_q(m, n) \leq m$, *then the fraction of information systems from* $I_k(m, n)$ *having minimal inhibitory rules with the length at most* $\varrho_q(m, n)$ *is at most*

$$\frac{1}{k^{(k^{q+1}/\ln k - 2) \log_k m \log_k n}} .$$

2.6.2 Number of Minimal Inhibitory Rules

In this subsection, we consider k-valued information systems with m attributes and $n = \lfloor m^\alpha \rfloor$ objects, where α is a positive real number. We obtain upper bound on the length of minimal inhibitory rules, and lower and upper bounds on the number $M(S)$ of minimal inhibitory rules for almost all information systems $S \in I_k(m, \lfloor m^\alpha \rfloor)$.

Next statement follows immediately from Theorem 2.6 and Corollary 2.2.

Theorem 2.9. *Let* α *be a positive real number. Then for almost all information systems* $S \in I_k(m, \lfloor m^\alpha \rfloor)$ *the length of each minimal inhibitory rule is at most*

$$\lceil \log_k m + \log_k \lfloor m^\alpha \rfloor + 3 \rceil - 1 ,$$

and

$$m^{(1/5)\alpha \log_k m} \leq M(S) \leq m^{2(1+\alpha) \log_k m + 7} .$$

2.6.3 Construction of All Minimal Inhibitory Rules

Let α be a positive real number. We consider k-valued information systems with m attributes and $n = \lfloor m^\alpha \rfloor$ objects. For a given information system S from $I_k(m, \lfloor m^\alpha \rfloor)$ it is required to construct all minimal inhibitory rules over S. The length of input for this problem is equal to $m \lfloor m^\alpha \rfloor \lceil \log_2 k \rceil \leq m^{1+\alpha} \lceil \log_2 k \rceil$. The length of output is at least $M(S)$.

Let $\tau(\alpha, m) = \lceil \log_k m + \log_k \lfloor m^\alpha \rfloor + 3 \rceil - 1$. From Theorem 2.9 it follows that for almost all information systems S from $I_k(m, \lfloor m^\alpha \rfloor)$ the length of each minimal inhibitory rule is at most $\tau(\alpha, m)$, and

$$m^{(1/5)\alpha \log_k m} \leq M(S) \leq m^{2(1+\alpha) \log_k m + 7} .$$

Thus, there is no algorithm which for almost all information systems from $I_k(m, \lfloor m^\alpha \rfloor)$ constructs the set of all minimal inhibitory rules and for such information systems has polynomial time complexity depending on the length of input.

Let us consider an algorithm which finds all inhibitory rules with the length at most $\tau = \tau(\alpha, m)$ and for each such rule recognizes if this rule is minimal inhibitory or not. It is clear that this recognition problem can be solved (for one rule) in polynomial time depending on the length of input. The number of such rules is at most $k^\tau m^\tau$. One can show that $k^\tau m^\tau \le m^{3(1+\alpha)\log_k m}$ for sufficiently large m. From Theorem 2.9 it follows that for almost all information systems from $I_k(m, \lfloor m^\alpha \rfloor)$ the considered algorithm finds all minimal inhibitory rules.

The considered algorithm works with at most $m^{3(1+\alpha)\log_k m}$ rules for sufficiently large m. Using Theorem 2.9 we conclude that for almost all information systems S from $I_k(m, \lfloor m^\alpha \rfloor)$,

$$m^{3(1+\alpha)\log_k m} = \left(m^{(1/5)\alpha \log_k m} \right)^{15(1+\alpha)/\alpha} \le (M(S))^{15(1+\alpha)/\alpha} \ .$$

Thus, there exists an algorithm which for almost all information systems from $I_k(m, \lfloor m^\alpha \rfloor)$ constructs the set of minimal inhibitory rules and for such information systems has polynomial time complexity depending on the length of input and the length of output.

2.7 Conclusions

In this chapter, we have studied k-valued information systems with polynomial number of objects in the number of attributes. Almost all information systems have relatively short minimal inhibitory association rules only but the size of the set of minimal inhibitory association rules for such information systems is not polynomial in the number of attributes. Analogous results are obtained for minimal inhibitory decision rules. Hence, we should consider heuristics for the generation of some relevant sets of inhibitory rules, and lazy algorithms which check some properties of inhibitory rules without their direct generation.

3 Partial Covers and Inhibitory Decision Rules

In this chapter, we consider algorithms for construction of partial inhibitory decision rules and some bounds on the length such rules. These investigations are based on the use of known results for partial covers.

We show that:

- Under some natural assumptions on the class NP, the greedy algorithm is close to the best polynomial approximate algorithms for the minimization of the length of partial inhibitory decision rules.
- Based on an information received during the greedy algorithm work, it is possible to obtain nontrivial lower and upper bounds on the minimal length of partial inhibitory decision rules.
- For the most part of randomly generated binary decision tables, greedy algorithm constructs simple partial inhibitory decision rules with relatively high accuracy. In particular, some theoretical results confirm the following 0.5-hypothesis for inhibitory decision rules: for the most part of decision tables for each row during each step the greedy algorithm chooses an attribute that separates from the considered row at least one-half of rows that should be separated.

Similar results can be obtained for partial inhibitory association rules over information systems. To this end, it is enough to fix an arbitrary attribute a_i of the information system as the decision attribute and study inhibitory association rules with the right-hand side of the kind $a_i \neq c$ as inhibitory decision rules over the obtained decision system (decision table).

In this chapter, we use the collection of results on partial covers mentioned or obtained in [47, 48] together with some definitions and examples.

The chapter consists of three sections. In Sect. 3.1, known results for partial covers are recalled. In Sect. 3.2, partial inhibitory decision rules are studied. Section 3.3 contains short conclusions.

3.1 Partial Covers

This section consists of six subsections. In Sect. 3.1.1, main notions are described. In Sect. 3.1.2, some known results on partial covers are recalled. In Sect. 3.1.3, polynomial approximate algorithms for partial cover minimization (construction

P. Delimata et al.: Inhibitory Rules in Data Analysis, SCI 163, pp. 43–62.
springerlink.com   © Springer-Verlag Berlin Heidelberg 2009

of partial cover with minimal cardinality) are discussed. In Sect. 3.1.4, upper and lower bounds on minimal cardinality of partial covers based on an information about the greedy algorithm work are described. In Sect. 3.1.5, an upper bound on cardinality of partial cover constructed by the greedy algorithm is presented. In Sect. 3.1.6, exact and partial covers for the most part of set cover problems are discussed.

3.1.1 Main Notions

Let $A = \{e_1, \ldots, e_n\}$ be a nonempty finite set and $S = \{B_i\}_{i \in \{1, \ldots, m\}} = \{B_1, \ldots, B_m\}$ be a family of subsets of A such that $B_1 \cup \ldots \cup B_m = A$. We assume that S can contain equal subsets of A. The pair (A, S) is called a *set cover problem*.

Let I be a subset of $\{1, \ldots, m\}$. The family $P = \{B_i\}_{i \in I}$ is called a *subfamily* of S. The number $|I|$ is called the *cardinality* of P and is denoted by $|P|$. Let $P = \{B_i\}_{i \in I}$ and $Q = \{B_i\}_{i \in J}$ be subfamilies of S. The notation $P \subseteq Q$ means that $I \subseteq J$. Let $P \cup Q = \{B_i\}_{i \in I \cup J}$, $P \cap Q = \{B_i\}_{i \in I \cap J}$, and $P \setminus Q = \{B_i\}_{i \in I \setminus J}$.

A subfamily $Q = \{B_{i_1}, \ldots, B_{i_t}\}$ of the family S is called a *partial cover for* (A, S). Let $\alpha \in \mathbb{R}$ and $0 \leq \alpha < 1$. The subfamily Q is called an *α-cover for* (A, S) if $|B_{i_1} \cup \ldots \cup B_{i_t}| \geq (1 - \alpha)|A|$. For example, 0.01-cover means that we should cover at least 99% of elements from A. Note that a 0-cover is an exact cover. By $C_{\min}(\alpha) = C_{\min}(\alpha, A, S)$ we denote the minimal cardinality of α-cover for (A, S). The notation $C_{\min}(\alpha)$ will be used in cases, where A and S are known.

Let us consider a greedy algorithm with threshold α (see Algorithm 3.1) which constructs an α-cover for (A, S).

Algorithm 3.1. [47] Greedy algorithm for partial cover construction

---

Input : Set cover problem (A, S) with $S = \{B_1, \ldots, B_m\}$, and real number α,
            $0 \leq \alpha < 1$.
Output: α-cover for (A, S).
$Q \longleftarrow \emptyset$;
while Q *is not an α-cover for* (A, S) **do**
    select $B_i \in S$ with minimal index i such that B_i covers the maximal number
    of elements from A uncovered by subsets from Q;
    $Q \longleftarrow Q \cup \{B_i\}$;
end
return Q;

By $C_{\text{greedy}}(\alpha) = C_{\text{greedy}}(\alpha, A, S)$ we denote the cardinality of constructed α-cover for (A, S).

3.1.2 Some Known Results

First, we consider some known results for exact covers, where $\alpha = 0$.

Theorem 3.1. (Nigmatullin [57])

$$C_{\text{greedy}}(0) \leq C_{\min}(0)(1 + \ln |A| - \ln C_{\min}(0)) .$$

Theorem 3.2. (Johnson [23], Lovász [28])

$$C_{\text{greedy}}(0) \leq C_{\min}(0)(1 + \ln(\max_{B_i \in S} |B_i|)) \leq C_{\min}(0)(1 + \ln |A|) .$$

More exact bounds (depending only on $|A|$) were obtained by Slavík [76, 77].

Theorem 3.3. (Slavík [76, 77]) *If* $|A| \geq 2$, *then*

$$C_{\text{greedy}}(0) < C_{\min}(0)(\ln |A| - \ln \ln |A| + 0.78) .$$

Theorem 3.4. (Slavík [76, 77]) *For any natural* $m \geq 2$, *there exists a set cover problem* (A, S) *such that* $|A| = m$ *and*

$$C_{\text{greedy}}(0) > C_{\min}(0)(\ln |A| - \ln \ln |A| - 0.31) .$$

There are some results on exact and approximate polynomial algorithms for cover minimization.

Theorem 3.5. (Karp [24]) *The problem of construction of 0-cover with minimal cardinality is* NP-*hard.*

Theorem 3.6. (Feige [18]) *If* $NP \not\subseteq DTIME(n^{O(\log \log n)})$, *then for any* ε, $0 < \varepsilon < 1$, *there is no polynomial algorithm that for a given set cover problem* (A, S) *constructs a 0-cover for* (A, S) *which cardinality is at most*

$$(1 - \varepsilon)C_{\min}(0) \ln |A| .$$

Theorem 3.7. (Raz and Safra [68]) *If* $P \neq NP$, *then there exists* $\gamma > 0$ *such that there is no polynomial algorithm that for a given set cover problem* (A, S) *constructs a 0-cover for* (A, S) *which cardinality is at most*

$$\gamma C_{\min}(0) \ln |A| .$$

We now consider some known results for partial covers, where $\alpha \geq 0$.

Theorem 3.8. (Slavík [76, 77]) *Let* $0 \leq \alpha < 1$ *and* $\lceil (1 - \alpha)|A| \rceil \geq 2$. *Then*

$$C_{\text{greedy}}(\alpha) < C_{\min}(\alpha)(\ln \lceil (1 - \alpha)|A| \rceil - \ln \ln \lceil (1 - \alpha)|A| \rceil + 0.78) .$$

Theorem 3.9. (Slavík [76, 77]) *Let* $0 \leq \alpha < 1$. *Then for any natural* $t \geq 2$ *there exists a set cover problem* (A, S) *such that* $\lceil (1 - \alpha)|A| \rceil = t$ *and*

$$C_{\text{greedy}}(\alpha) > C_{\min}(\alpha)(\ln \lceil (1 - \alpha)|A| \rceil - \ln \ln \lceil (1 - \alpha)|A| \rceil - 0.31) .$$

Theorem 3.10. (Slavík [77]) *Let* $0 \leq \alpha < 1$. *Then*

$$C_{\text{greedy}}(\alpha) \leq C_{\min}(\alpha)(1 + \ln(\max_{B_i \in S} |B_i|)) .$$

There are some bounds on $C_{\mathrm{greedy}}(\alpha)$ which does not depend on $|A|$. Note that in the next two theorems we consider the case, where $\alpha > 0$.

Theorem 3.11. (Cheriyan and Ravi [10]) *Let $0 < \alpha < 1$. Then*

$$C_{\mathrm{greedy}}(\alpha) \leq C_{\mathrm{min}}(0) \ln(1/\alpha) + 1 .$$

This bound was rediscovered by Moshkov in [34] and generalized in [35].

Theorem 3.12. (Moshkov [35]) *Let $0 < \beta \leq \alpha < 1$. Then*

$$C_{\mathrm{greedy}}(\alpha) \leq C_{\mathrm{min}}(\alpha - \beta) \ln(1/\beta) + 1 .$$

There is a result on exact polynomial algorithms for partial cover minimization.

Theorem 3.13. (Ślęzak [79, 81]) *Let $0 \leq \alpha < 1$. Then the problem of construction of α-cover with minimal cardinality is NP-hard.*

3.1.3   Polynomial Approximate Algorithms

In [47], using technique created by Ślęzak in [79, 81], Moshkov, Piliszczuk and Zielosko generalized the results of Feige, Raz and Safra (Theorems 3.6 and 3.7) to the case of partial covers.

Theorem 3.14. (Moshkov, Piliszczuk, Zielosko [47]) *Let $\alpha \in \mathbb{R}$ and $0 \leq \alpha < 1$. If $NP \not\subseteq DTIME(n^{O(\log \log n)})$, then for any ε, $0 < \varepsilon < 1$, there is no polynomial algorithm that for a given set cover problem (A, S) constructs an α-cover for (A, S) which cardinality is at most*

$$(1 - \varepsilon)C_{\mathrm{min}}(\alpha, A, S) \ln|A| .$$

From Theorem 3.10 it follows that $C_{\mathrm{greedy}}(\alpha) \leq C_{\mathrm{min}}(\alpha)(1 + \ln|A|)$. From this inequality and from Theorem 3.14 it follows that, under the assumption $NP \not\subseteq DTIME(n^{O(\log \log n)})$, the greedy algorithm is close to the best polynomial approximate algorithms for partial cover minimization.

Theorem 3.15. (Moshkov, Piliszczuk, Zielosko [47]) *Let $\alpha \in \mathbb{R}$ and $0 \leq \alpha < 1$. If $P \neq NP$, then there exists $\varrho > 0$ such that there is no polynomial algorithm that for a given set cover problem (A, S) constructs an α-cover for (A, S) which cardinality is at most*

$$\varrho C_{\mathrm{min}}(\alpha, A, S) \ln|A| .$$

3.1.4   Bounds on $C_{\mathrm{min}}(\alpha)$ Based on Information About Greedy Algorithm Work

In this subsection, we fix some information on the greedy algorithm work, and consider the best upper and lower bounds on $C_{\mathrm{min}}(\alpha)$ depending on this information. These bounds were obtained in [47].

Information on Greedy Algorithm Work

Let us assume that (A, S) is a set cover problem and α is a real number such that $0 \leq \alpha < 1$. We now apply the greedy algorithm with threshold α to the problem (A, S). Let us assume that during the construction of α-cover the greedy algorithm chooses consequently subsets $B_{j_1}, \ldots, B_{j_t}$. Set $B_{j_0} = \emptyset$ and for $i = 1, \ldots, t$ set $\delta_i = |B_{j_i} \setminus (B_{j_0} \cup \ldots \cup B_{j_{i-1}})|$.

Let us define $\Delta(\alpha, A, S) = (\delta_1, \ldots, \delta_t)$. As information on the greedy algorithm work we will use the tuple $\Delta(\alpha, A, S)$ and numbers $|A|$ and α. Note that $\delta_1 = \max\{|B_i| : B_i \in S\}$ and $t = C_{\text{greedy}}(\alpha, A, S)$. Let us denote by P_{SC} the set of set cover problems and $D_{SC} = \{(\alpha, |A|, \Delta(\alpha, A, S)) : \alpha \in \mathbb{R}, 0 \leq \alpha < 1, (A, S) \in P_{SC}\}$.

Lemma 3.16. (Moshkov, Piliszczuk, Zielosko [47]) *A tuple $(\alpha, n, (\delta_1, \ldots, \delta_t))$ belongs to the set D_{SC} if and only if α is a real number such that $0 \leq \alpha < 1$, and $n, \delta_1, \ldots, \delta_t$ are natural numbers such that $\delta_1 \geq \ldots \geq \delta_t$, $\sum_{i=1}^{t-1} \delta_i < (1-\alpha)n$ and $(1 - \alpha)n \leq \sum_{i=1}^{t} \delta_i \leq n$.*

The Best Upper Bound on $C_{\min}(\alpha)$

We define a function $\mathcal{U}_{SC} : D_{SC} \rightarrow \mathbb{N}$. Let $(\alpha, n, (\delta_1, \ldots, \delta_t)) \in D_{SC}$. Then $\mathcal{U}_{SC}(\alpha, n, (\delta_1, \ldots, \delta_t)) = \max\{C_{\min}(\alpha, A, S) : (A, S) \in P_{SC}, |A| = n, \Delta(\alpha, A, S) = (\delta_1, \ldots, \delta_t)\}$. It is clear that

$$C_{\min}(\alpha, A, S) \leq \mathcal{U}_{SC}(\alpha, |A|, \Delta(\alpha, A, S))$$

is the best upper bound on $C_{\min}(\alpha)$ depending on α, $|A|$ and $\Delta(\alpha, A, S)$.

Theorem 3.17. (Moshkov, Piliszczuk, Zielosko [47]) *Let $(\alpha, n, (\delta_1, \ldots, \delta_t)) \in D_{SC}$. Then*

$$\mathcal{U}_{SC}(\alpha, n, (\delta_1, \ldots, \delta_t)) = t .$$

Thus, $C_{\min}(\alpha, A, S) \leq C_{\text{greedy}}(\alpha, A, S)$ is the best upper bound on $C_{\min}(\alpha)$ depending on α, $|A|$ and $\Delta(\alpha, A, S)$.

The Best Lower Bound on $C_{\min}(\alpha)$

We define a function $\mathcal{L}_{SC} : D_{SC} \rightarrow \mathbb{N}$. Let $(\alpha, n, (\delta_1, \ldots, \delta_t)) \in D_{SC}$. Then $\mathcal{L}_{SC}(\alpha, n, (\delta_1, \ldots, \delta_t)) = \min\{C_{\min}(\alpha, A, S) : (A, S) \in P_{SC}, |A| = n, \Delta(\alpha, A, S) = (\delta_1, \ldots, \delta_t)\}$. It is clear that

$$C_{\min}(\alpha, A, S) \geq \mathcal{L}_{SC}(\alpha, |A|, \Delta(\alpha, A, S))$$

is the best lower bound on $C_{\min}(\alpha)$ depending on α, $|A|$ and $\Delta(\alpha, A, S)$. For $(\alpha, n, (\delta_1, \ldots, \delta_t)) \in D_{SC}$ and $\delta_0 = 0$, set

$$l(\alpha, n, (\delta_1, \ldots, \delta_t)) = \max\left\{ \left\lceil \frac{\lceil (1-\alpha)n \rceil - (\delta_0 + \ldots + \delta_i)}{\delta_{i+1}} \right\rceil : i = 0, \ldots, t - 1 \right\}.$$

Theorem 3.18. (Moshkov, Piliszczuk, Zielosko [47]) *Let* $(\alpha, n, (\delta_1, \ldots, \delta_t)) \in D_{SC}$. *Then*

$$\mathcal{L}_{SC}(\alpha, n, (\delta_1, \ldots, \delta_t)) = l(\alpha, n, (\delta_1, \ldots, \delta_t)) .$$

So $C_{\min}(\alpha, A, S) \geq l(\alpha, |A|, \Delta(\alpha, A, S))$ is the best lower bound on $C_{\min}(\alpha)$ depending on α, $|A|$ and $\Delta(\alpha, A, S)$.

Properties of the Best Lower Bound on $C_{\min}(\alpha)$

Let us assume that (A, S) is a set cover problem and α is a real number such that $0 \leq \alpha < 1$. Let

$$l_{SC}(\alpha) = l_{SC}(\alpha, A, S) = l(\alpha, |A|, \Delta(\alpha, A, S)) .$$

Lemma 3.19. (Moshkov, Piliszczuk, Zielosko [47]) *Let* $\alpha_1, \alpha_2 \in \mathbb{R}$ *and* $0 \leq \alpha_1 < \alpha_2 < 1$. *Then*

$$l_{SC}(\alpha_1) \geq l_{SC}(\alpha_2) .$$

Corollary 3.20. (Moshkov, Piliszczuk, Zielosko [47])

$$l_{SC}(0) = \max\{l_{SC}(\alpha) : 0 \leq \alpha < 1\} .$$

The value $l_{SC}(\alpha)$ can be used for obtaining upper bounds on the cardinality of partial covers constructed by the greedy algorithm.

Theorem 3.21. (Moshkov, Piliszczuk, Zielosko [47]) *Let* α *and* β *be real numbers such that* $0 < \beta \leq \alpha < 1$. *Then*

$$C_{\text{greedy}}(\alpha) < l_{SC}(\alpha - \beta) \ln \left(\frac{1 - \alpha + \beta}{\beta} \right) + 1 .$$

Corollary 3.22. (Moshkov, Piliszczuk, Zielosko [47]) *Let* $\alpha \in \mathbb{R}$, $0 < \alpha < 1$. *Then*

$$C_{\text{greedy}}(\alpha) < l_{SC}(0) \ln \left(\frac{1}{\alpha} \right) + 1 .$$

If $l_{SC}(0)$ is a small number, then we have a good upper bound on $C_{\text{greedy}}(\alpha)$. If $l_{SC}(0)$ is a big number, then we have a big lower bound on $C_{\min}(0)$ and on $C_{\min}(\alpha)$ for some α.

3.1.5   Upper Bound on $C_{\text{greedy}}(\alpha)$

In this subsection, we consider one more upper bound on $C_{\text{greedy}}(\alpha)$ which does not depend on $|A|$. In some sense, this bound is unimprovable.

Theorem 3.23. (Moshkov, Piliszczuk, Zielosko [47]) *Let* α *and* β *be real numbers such that* $0 < \beta \leq \alpha < 1$. *Then*

$$C_{\text{greedy}}(\alpha) < C_{\min}(\alpha - \beta) \ln \left(\frac{1 - \alpha + \beta}{\beta} \right) + 1 .$$

Theorem 3.24. (Moshkov, Piliszczuk, Zielosko [47]) *There is no real $\delta < 1$ such that for any set cover problem (A, S) and for any real α and β, $0 < \beta \leq \alpha < 1$, the following inequality holds:*

$$C_{\text{greedy}}(\alpha) \leq \delta \left(C_{\min}(\alpha - \beta) \ln \left(\frac{1 - \alpha + \beta}{\beta} \right) + 1 \right) .$$

3.1.6   Covers for the Most Part of Set Cover Problems

In this subsection, covers for the most part of set cover problem are discussed.

Exact Covers for the Most Part of Set Cover Problems

First, we consider exact covers for the most part of set cover problems such that $m \geq \lceil \log_2 n \rceil + t$ and t is large enough.

Theorem 3.25. (Moshkov, Piliszczuk, Zielosko [47]) *Let us consider set cover problems (A, S) such that $A = \{e_1, \dots, e_n\}$, $S = \{B_1, \dots, B_m\}$ and $m \geq \lceil \log_2 n \rceil + t$, where t is a natural number. Let $i_1, \dots, i_{\lceil \log_2 n \rceil + t}$ be pairwise different numbers from $\{1, \dots, m\}$. Then the fraction of set cover problems (A, S), for which $\{B_{i_1}, \dots, B_{i_{\lceil \log_2 n \rceil + t}}\}$ is an exact cover for (A, S), is at least*

$$1 - \frac{1}{2^t - 1} .$$

For example, if $t = 7$, then for at least 99% of set cover problems (A, S) the subsets $B_{i_1}, \dots, B_{i_{\lceil \log_2 n \rceil + t}}$ form an exact cover for (A, S).

So if $m \geq \lceil \log_2 n \rceil + t$ and t is large enough, then for the most part of set cover problems there exist exact (and, consequently, partial) covers with small cardinality.

Partial Covers Constructed by Greedy Algorithm for the Most Part of Set Cover Problems

We now consider the behavior of greedy algorithm for the most part of set cover problems such that $m \geq n + t$ and t is large enough.

Let us consider set cover problems (A, S) such that $A = \{e_1, \dots, e_n\}$ and $S = \{B_1, \dots, B_m\}$. A problem (A, S) will be called *saturated* if for any nonempty subset A' of A there exists a subset B_i from S which covers at least one-half of elements from A'. For a saturated set cover problem, the greedy algorithm at each step chooses a subset which covers at least one-half of uncovered elements. So for saturated set cover problems the following 0.5-hypothesis, formulated in [48], is true: for the most part of set cover problems, during each step the greedy algorithm chooses a subset which covers at least one-half of uncovered elements.

Theorem 3.26. (Moshkov, Piliszczuk, Zielosko [47]) *Let us consider set cover problems* (A, S) *such that* $A = \{e_1, \ldots, e_n\}$, $S = \{B_1, \ldots, B_m\}$ *and* $m > n$. *Then the fraction of saturated set cover problems* (A, S) *is at least*

$$1 - \frac{1}{2^{m-n} - 1} \, .$$

For example, if $m = n + 7$, then at least 99% of set cover problems are saturated.

We now repeat simple reasoning from [47]. Let us analyze the work of greedy algorithm on an arbitrary saturated set cover problem (A, S). For $i = 1, 2, \ldots$, after the step number i at most $|A|/2^i$ elements from A are uncovered. We now evaluate the number $C_{\text{greedy}}(\alpha)$, where $0 < \alpha < 1$. It is clear that $C_{\text{greedy}}(\alpha) \leq i$, where i is a number such that $1/2^i \leq \alpha$. One can show that $1/2^{\lceil \log_2(1/\alpha) \rceil} \leq \alpha$. Therefore, $C_{\text{greedy}}(\alpha) \leq \lceil \log_2(1/\alpha) \rceil$. Some examples can be found in Table 3.1.

Table 3.1. [47] Values of $\lceil \log_2(1/\alpha) \rceil$ for some α

α	0.5	0.3	0.1	0.01	0.001
Percentage of covered elements	50	70	90	99	99.9
$\lceil \log_2(1/\alpha) \rceil$	1	2	4	7	10

Let us evaluate the number $C_{\text{greedy}}(0)$. It is clear that all elements from A will be covered after a step number i if $|A|/2^i < 1$, i.e., if $i > \log_2 |A|$. If $\log_2 |A|$ is an integer, we can set $i = \log_2 |A| + 1$. Otherwise, we can set $i = \lceil \log_2 |A| \rceil$. Therefore, $C_{\text{greedy}}(0) \leq \log_2 |A| + 1$.

We now evaluate the number $l_{SC}(0)$. Let $\Delta(0, A, S) = (\delta_1, \ldots, \delta_m)$. Set $\delta_0 = 0$. Then $l_{SC}(0) = \max\{\lceil (|A| - (\delta_0 + \ldots + \delta_i))/\delta_{i+1} \rceil : i = 0, \ldots, m-1\}$. Since (A, S) is a saturated problem, we have $\delta_{i+1} \geq (|A| - (\delta_0 + \ldots + \delta_i))/2$ and $2 \geq (|A| - (\delta_0 + \ldots + \delta_i))/\delta_{i+1}$ for $i = 0, \ldots, m-1$. Therefore, $l_{SC}(0) \leq 2$. Using Corollary 3.20 we obtain $l_{SC}(\alpha) \leq 2$ for any α, $0 \leq \alpha < 1$.

Experimental results considered in [48] show that for the most part of randomly generated set cover problems (not only for the case, where $|S| > |A|$) during each step the greedy algorithm chooses a subset which covers at least one-half of uncovered elements.

3.2   Partial Inhibitory Decision Rules

This section consists of seven subsections. In Sect. 3.2.1, main notions are described. In Sect. 3.2.2, relationships between partial covers and partial inhibitory decision rules are presented. In Sect. 3.2.3, generalizations of Slavík's results to the case of partial inhibitory decision rules are given. In Sect. 3.2.4, polynomial approximate algorithms for partial inhibitory decision rule minimization (construction of partial inhibitory decision rule with minimal length) are studied. In Sect. 3.2.5, upper and lower bounds on minimal length of partial inhibitory decision rules based on an information about greedy algorithm work are investigated.

In Sect. 3.2.6, an upper bound on the length of partial inhibitory decision rule constructed by greedy algorithm is considered. In Sect. 3.2.7, inhibitory decision rules for the most part of binary decision tables are discussed.

3.2.1   Main Notions

We assume that T is a decision table with n rows labeled with nonnegative integers (values of the decision attribute d) and m columns labeled with conditional attributes (names of attributes) $a_1, \ldots, a_m$. This table is filled by nonnegative integers (values of conditional attributes). We denote by $Dec(T)$ the set of decisions attached to rows of T. We will assume later that $|Dec(T)| \geq 2$.

Let $r = (b_1, \ldots, b_m)$ be a row of T labeled with a decision b, and c be a decision from $Dec(T)$ such that $c \neq b$. By $U(T, r, c)$ we denote the set of rows from T which are different (in at least one column) from r and are labeled with the decision c. We will say that an attribute a_i *separates* a row $r' \in U(T, r, c)$ from the row r if the rows r and r' have different numbers at the intersection with column a_i. The triple (T, r, c) will be called an *inhibitory decision rule problem*.

Let $0 \leq \alpha < 1$. A rule

$$(a_{i_1} = b_{i_1}) \wedge \ldots \wedge (a_{i_t} = b_{i_t}) \to d \neq c \tag{3.1}$$

is called an α-*inhibitory decision rule* for (T, r, c) if attributes $a_{i_1}, \ldots, a_{i_t}$ separate from r at least $(1 - \alpha)|U(T, r, c)|$ rows from $U(T, r, c)$ (such rules are also called *partial* inhibitory decision rules). The number t is called the *length* of the considered inhibitory decision rule. If $U(T, r, c) = \emptyset$, then for any $a_{i_1}, \ldots, a_{i_t} \in \{a_1, \ldots, a_m\}$ the rule (3.1) is an α-inhibitory decision rule for (T, r, c). The rule (3.1) with empty left-hand side (when $t = 0$) is also an α-inhibitory decision rule for (T, r, c).

For example, 0.01-decision rule means that we should separate from r at least 99% of rows from $U(T, r, c)$. Note that a 0-inhibitory decision rule is an exact inhibitory decision rule. By $L_{\min}(\alpha) = L_{\min}(\alpha, T, r, c)$ we denote the minimal length of α-inhibitory decision rule for (T, r, c).

We now describe a greedy algorithm with threshold α which constructs an α-inhibitory decision rule for (T, r, c) (see Algorithm 3.2).

Let us denote by $L_{\text{greedy}}(\alpha) = L_{\text{greedy}}(\alpha, T, r, c)$ the length of constructed α-inhibitory decision rule for (T, r, c).

3.2.2   Relationships between Partial Covers and Partial Inhibitory Decision Rules

Let T be a decision table with m columns labeled with attributes $a_1, \ldots, a_m$, r be a row from T labeled with the decision b, $c \in Dec(T)$, $c \neq b$, and $U(T, r, c)$ be a nonempty set.

We correspond a set cover problem $(A(T, r, c), S(T, r, c))$ to the considered inhibitory decision rule problem (T, r, c) in the following way: $A(T, r, c) = U(T, r, c)$ and $S(T, r, c) = \{B_1, \ldots, B_m\}$, where $B_1 = U(T, r, c, a_1), \ldots, B_m =$

Algorithm 3.2. Greedy algorithm for partial inhibitory decision rule construction

---

Input : Decision table T with conditional attributes $a_1, \ldots, a_m$, row $r = (b_1, \ldots, b_m)$ of T labeled with the decision b, decision $c \in Dec(T)$ such that $c \neq b$ and real number α, $0 \leq \alpha < 1$.

Output: α-inhibitory decision rule for (T, r, c).

$Q \longleftarrow \emptyset$;

while *attributes from Q separate from r less than* $(1 - \alpha)|U(T, r, c)|$ *rows from* $U(T, r, c)$ **do**

select $a_i \in \{a_1, \ldots, a_m\}$ with minimal index i such that a_i separates from r the maximal number of rows from $U(T, r, c)$ unseparated by attributes from Q;

$Q \longleftarrow Q \cup \{a_i\}$;

end

return $\bigwedge_{a_i \in Q} (a_i = b_i) \rightarrow d \neq c$;

$U(T, r, c, a_m)$ and for $i = 1, \ldots, m$ the set $U(T, r, c, a_i)$ coincides with the set of rows from $U(T, r, c)$ separated by the attribute a_i from the row r.

Let during the construction of an α-inhibitory decision rule for (T, r, c) the greedy algorithm choose consequently attributes $a_{j_1}, \ldots, a_{j_t}$. Set

$$U(T, r, c, a_{j_0}) = \emptyset$$

and for $i = 1, \ldots, t$ set

$$\delta_i = |U(T, r, c, a_{j_i}) \setminus (U(T, r, c, a_{j_0}) \cup \ldots \cup U(T, r, c, a_{j_{i-1}}))| .$$

Let $\Delta(\alpha, T, c, r) = (\delta_1, \ldots, \delta_t)$. It is not difficult to prove the following statement.

Proposition 3.27. *Let α be a real number such that $0 \leq \alpha < 1$. Then $|U(T, r, c)| = |A(T, r, c)|$, $\Delta(\alpha, T, r, c) = \Delta(\alpha, A(T, r, c), S(T, r, c))$, and*

$$L_{\min}(\alpha, T, r, c) = C_{\min}(\alpha, A(T, r, c), S(T, r, c)) ,$$
$$L_{\text{greedy}}(\alpha, T, r, c) = C_{\text{greedy}}(\alpha, A(T, r, c), S(T, r, c)) .$$

Let (A, S) be a set cover problem, $A = \{e_1, \ldots, e_n\}$ and $S = \{B_1, \ldots, B_m\}$. We correspond an inhibitory decision rule problem $(T(A, S), r(A, S), c(A, S))$ to the set cover problem (A, S) in the following way. The table $T(A, S)$ contains m columns labeled with attributes $a_1, \ldots, a_m$ and $n + 1$ rows filled by numbers from $\{0, 1\}$. For $i = 1, \ldots, n$ and $j = 1, \ldots, m$, the number 1 stays at the intersection of i-th row and j-th column if and only if $e_i \in B_j$. The $(n + 1)$-th row is filled by 0. Let us consider the values of the decision attribute d. The first n rows are labeled with the decision 0. The last row is labeled with the decision 1. Let us denote by $r(A, S)$ the last row of the table $T(A, S)$. By $c(A, S)$ we denote the decision 0. For $i \in \{1, \ldots, n + 1\}$, we denote by r_i the i-th row. It is not difficult to see that $U(T(A, S), r(A, S), c(A, S)) = \{r_1, \ldots, r_n\}$. Let $i \in \{1, \ldots, n\}$ and $j \in \{1, \ldots, m\}$. One can show that the attribute a_j separates the row $r_{n+1} = r(A, S)$ from the row r_i if and only if $e_i \in B_j$. It is not difficult to prove the following statements.

Proposition 3.28. *Let $\alpha \in \mathbb{R}$, $0 \le \alpha < 1$, and $\{i_1, \ldots, i_t\} \subseteq \{1, \ldots, m\}$. Then*

$$(a_{i_1} = 0) \wedge \ldots \wedge (a_{i_t} = 0) \to d \neq 0$$

is an α-inhibitory decision rule for $(T(A, S), r(A, S), c(A, S))$ if and only if $\{B_{i_1}, \ldots, B_{i_t}\}$ is an α-cover for (A, S).

Proposition 3.29. *Let $\alpha \in \mathbb{R}$ and $0 \le \alpha < 1$. Then*

$$|U(T(A, S), r(A, S), c(A, S))| = |A| \, ,$$
$$L_{\min}(\alpha, T(A, S), r(A, S), c(A, S)) = C_{\min}(\alpha, A, S) \, ,$$
$$L_{\text{greedy}}(\alpha, T(A, S), r(A, S), c(A, S)) = C_{\text{greedy}}(\alpha, A, S) \, ,$$
$$\Delta(\alpha, T(A, S), r(A, S), c(A, S)) = \Delta(\alpha, A, S) \, .$$

Proposition 3.30. *There exists a polynomial algorithm which for a given set cover problem (A, S) constructs the inhibitory decision rule problem*

$$(T(A, S), r(A, S), c(A, S)) \, .$$

3.2.3 Precision of Greedy Algorithm

The following three statements are simple corollaries of results of Slavík (see Theorems 3.8–3.10). Let T be a decision table with m columns labeled with attributes $a_1, \ldots, a_m$, r be a row of T labeled with the decision b, $c \in Dec(T)$ and $c \neq b$.

Theorem 3.31. *Let $0 \le \alpha < 1$ and $\lceil (1 - \alpha)|U(T, r, c)| \rceil \ge 2$. Then*

$$L_{\text{greedy}}(\alpha) < L_{\min}(\alpha)(\ln \lceil (1 - \alpha)|U(T, r, c)| \rceil - \ln \ln \lceil (1 - \alpha)|U(T, r, c)| \rceil$$
$$+ 0.78) \, .$$

Proof. Let us denote $(A, S) = (A(T, r, c), S(T, r, c))$. From Proposition 3.27 it follows that $|A| = |U(T, r, c)|$. Therefore, $\lceil (1 - \alpha)|A| \rceil \ge 2$. Using Theorem 3.8 we obtain

$$C_{\text{greedy}}(\alpha, A, S) < C_{\min}(\alpha, A, S)(\ln \lceil (1 - \alpha)|A| \rceil - \ln \ln \lceil (1 - \alpha)|A| \rceil + 0.78) \, .$$

Using Proposition 3.27 we conclude that $L_{\text{greedy}}(\alpha) = C_{\text{greedy}}(\alpha, A, S)$ and $L_{\min}(\alpha) = C_{\min}(\alpha, A, S)$. Taking into account that $|A| = |U(T, r, c)|$ we conclude that the statement of the theorem holds. $\square$

Theorem 3.32. *Let $0 \le \alpha < 1$. Then for any natural $t \ge 2$ there exists an inhibitory decision rule problem (T, r, c) such that $\lceil (1 - \alpha)|U(T, r, c)| \rceil = t$ and*

$$L_{\text{greedy}}(\alpha) > L_{\min}(\alpha)(\ln \lceil (1 - \alpha)|U(T, r, c)| \rceil - \ln \ln \lceil (1 - \alpha)|U(T, r, c)| \rceil$$
$$- 0.31) \, .$$

Proof. From Theorem 3.9 it follows that for any natural $t \geq 2$ there exists a set cover problem (A, S) such that $\lceil (1 - \alpha)|A| \rceil = t$ and $C_{\text{greedy}}(\alpha, A, S) > C_{\min}(\alpha, A, S)(\ln \lceil (1 - \alpha)|A| \rceil - \ln \ln \lceil (1 - \alpha)|A| \rceil - 0.31)$.

Let us consider the problem $(T, r, c) = (T(A, S), r(A, S), c(A, S))$. By Proposition 3.29, $|U(T, r, c)| = |A|$, $C_{\text{greedy}}(\alpha, A, S) = L_{\text{greedy}}(\alpha, T, r, c)$ and $C_{\min}(\alpha, A, S) = L_{\min}(\alpha, T, r, c)$. Hence, the statement of the theorem holds. $\square$

Theorem 3.33. *Let* $0 \leq \alpha < 1$ *and* $U(T, r, c) \neq \emptyset$. *Then*

$$L_{\text{greedy}}(\alpha) \leq L_{\min}(\alpha)(1 + \ln(\max_{j \in \{1, \ldots, m\}} |U(T, r, c, a_j)|)) .$$

Proof. Let us consider the set cover problem $(A, S) = (A(T, r, c), S(T, r, c))$. The inequality

$$C_{\text{greedy}}(\alpha, A, S) \leq C_{\min}(\alpha, A, S)(1 + \ln(\max_{j \in \{1, \ldots, m\}} |U(T, r, c, a_j)|))$$

follows from Theorem 3.10.

Using Proposition 3.27 we conclude that $C_{\text{greedy}}(\alpha, A, S) = L_{\text{greedy}}(\alpha)$ and $C_{\min}(\alpha, A, S) = L_{\min}(\alpha)$. Therefore, the statement of the theorem holds. $\square$

3.2.4   Polynomial Approximate Algorithms

Theorem 3.34. *Let* $0 \leq \alpha < 1$. *Then the problem of construction of α-inhibitory decision rule with minimal length is NP-hard.*

Proof. From Theorem 3.13 it follows that the problem of construction of α-cover with minimal cardinality is NP-hard. Using Propositions 3.28 and 3.30 we conclude that there exists a polynomial-time reduction of the problem of construction of α-cover with minimal cardinality to the problem of construction of α-inhibitory decision rule with minimal length. $\square$

Let us generalize Theorem 3.14 to the case of partial inhibitory decision rules.

Theorem 3.35. *Let* $\alpha \in \mathbb{R}$ *and* $0 \leq \alpha < 1$. *If* $NP \not\subseteq DTIME(n^{O(\log \log n)})$, *then for any* ε, $0 < \varepsilon < 1$, *there is no polynomial algorithm that for a given inhibitory decision rule problem* (T, r, c) *with* $U(T, r, c) \neq \emptyset$ *constructs an α-inhibitory decision rule for* (T, r, c) *which length is at most*

$$(1 - \varepsilon)L_{\min}(\alpha, T, r, c) \ln |U(T, r, c)| .$$

Proof. We assume the contrary: let $NP \not\subseteq DTIME(n^{O(\log \log n)})$ and for some ε, $0 < \varepsilon < 1$, a polynomial algorithm $\mathcal{A}$ exist that for a given inhibitory decision rule problem (T, r, c) with $U(T, r, c) \neq \emptyset$ constructs an α-inhibitory decision rule for (T, r, c) which length is at most $(1 - \varepsilon)L_{\min}(\alpha, T, r, c) \ln |U(T, r, c)|$.

Let (A, S) be an arbitrary set cover problem, $A = \{e_1, \ldots, e_n\}$ and $S = \{B_1, \ldots, B_m\}$. From Proposition 3.30 it follows that there exists a polynomial algorithm which for a given set cover problem (A, S) constructs the inhibitory decision rule problem $(T(A, S), r(A, S), c(A, S))$. Let us apply this algorithm and construct the inhibitory decision rule problem $(T, r, c) = (T(A, S), r(A, S), c(A, S))$. Let us apply to the inhibitory decision rule problem (T, r, c) the algorithm $\mathcal{A}$. As a result we obtain an α-inhibitory decision rule

$$(a_{i_1} = 0) \wedge \ldots \wedge (a_{i_t} = 0) \rightarrow d \neq 0$$

for (T, r, c) such that $t \leq (1 - \varepsilon) L_{\min}(\alpha, T, r, c) \ln |U(T, r, c)|$. From Proposition 3.28 it follows that $\{B_{i_1}, \ldots, B_{i_t}\}$ is an α-cover for (A, S). Using Proposition 3.29 we obtain $|A| = |U(T, r, c)|$ and $L_{\min}(\alpha, T, r, c) = C_{\min}(\alpha, A, S)$. Therefore, $t \leq (1 - \varepsilon) C_{\min}(\alpha, A, S) \ln |A|$.

Thus, under the assumption $NP \not\subseteq DTIME(n^{O(\log \log n)})$, there exists a polynomial algorithm that for a given set cover problem (A, S) constructs an α-cover for (A, S) which cardinality is at most $(1 - \varepsilon) C_{\min}(\alpha, A, S) \ln |A|$, but this fact contradicts Theorem 3.14. $\qquad \square$

From Theorem 3.33 it follows that $L_{\text{greedy}}(\alpha) \leq L_{\min}(\alpha)(1 + \ln |U(T, r, c)|)$. From this inequality and from Theorem 3.35 it follows that, under the assumption $NP \not\subseteq DTIME(n^{O(\log \log n)})$, the greedy algorithm is close to the best polynomial approximate algorithms for partial inhibitory decision rule minimization.

Let us generalize Theorem 3.15 to the case of partial inhibitory decision rules.

Theorem 3.36. *Let α be a real number such that $0 \leq \alpha < 1$. If $P \neq NP$, then there exists $\varrho > 0$ such that there is no polynomial algorithm that for a given inhibitory decision rule problem (T, r, c) with $U(T, r, c) \neq \emptyset$ constructs an α-inhibitory decision rule for (T, r, c) which length is at most*

$$\varrho L_{\min}(\alpha, T, r, c) \ln |U(T, r, c)| .$$

Proof. We now show that in the capacity of such ϱ we can choose ϱ from Theorem 3.15. Let us assume that the considered statement does not hold: let $P \neq NP$ and a polynomial algorithm $\mathcal{A}$ exist that for a given inhibitory decision rule problem (T, r, c) with $U(T, r, c) \neq \emptyset$ constructs an α-inhibitory decision rule for (T, r, c) which length is at most $\varrho L_{\min}(\alpha, T, r, c) \ln |U(T, r, c)|$.

Let (A, S) be an arbitrary set cover problem, $A = \{e_1, \ldots, e_n\}$ and $S = \{B_1, \ldots, B_m\}$. From Proposition 3.30 it follows that there exists a polynomial algorithm which for a given set cover problem (A, S) constructs the inhibitory decision rule problem $(T(A, S), r(A, S), c(A, S))$. Let us apply this algorithm and construct the inhibitory decision rule problem $(T, r, c) = (T(A, S), r(A, S), c(A, S))$. Let us apply to the problem (T, r, c) the algorithm $\mathcal{A}$. As a result we obtain an α-inhibitory decision rule

$$(a_{i_1} = 0) \wedge \ldots \wedge (a_{i_t} = 0) \rightarrow d \neq 0$$

for (T, r, c) such that $t \leq \varrho L_{\min}(\alpha, T, r, c) \ln |U(T, r, c)|$. From Proposition 3.28 it follows that $\{B_{i_1}, \ldots, B_{i_t}\}$ is an α-cover for (A, S). Using Proposition 3.29 we obtain $|A| = |U(T, r, c)|$ and $L_{\min}(\alpha, T, r, c) = C_{\min}(\alpha, A, S)$. Therefore, $t \leq \varrho C_{\min}(\alpha, A, S) \ln |A|$.

Thus, under the assumption $P \neq NP$, there exists a polynomial algorithm that for a given set cover problem (A, S) constructs an α-cover for (A, S) which cardinality is at most $\varrho C_{\min}(\alpha, A, S) \ln |A|$, but this fact contradicts Theorem 3.15. $\qquad \square$

3.2.5 Bounds on $L_{\min}(\alpha)$ Based on Information About Greedy Algorithm Work

In this subsection, we fix some information on the greedy algorithm work and find the best upper and lower bounds on $L_{\min}(\alpha)$ depending on this information.

Information on Greedy Algorithm Work

We assume that (T, r, c) is an inhibitory decision rule problem, where T is a decision table with m columns labeled with attributes $a_1, \ldots, a_m$, $U(T, r, c) \neq \emptyset$, and α is a real number such that $0 \leq \alpha < 1$. Let us apply the greedy algorithm with threshold α to the problem (T, r, c). Let during the construction of α-inhibitory decision rule the greedy algorithm choose consequently attributes $a_{j_1}, \ldots, a_{j_t}$. Set $U(T, r, c, a_{j_0}) = \emptyset$ and for $i = 1, \ldots, t$ set $\delta_i = |U(T, r, c, a_{j_i}) \setminus (U(T, r, c, a_{j_0}) \cup \ldots \cup U(T, r, c, a_{j_{i-1}}))|$. Let $\Delta(\alpha, T, r, c) = (\delta_1, \ldots, \delta_t)$. As information on the greedy algorithm work we will use the tuple $\Delta(\alpha, T, r, c)$, and numbers $|U(T, r, c)|$ and α. Note that $\delta_1 = \max\{|U(T, r, c, a_i)| : i = 1, \ldots, m\}$ and $t = L_{\text{greedy}}(\alpha, T, r, c)$.

Let us denote by P_{IDR} the set of inhibitory decision rule problems (T, r, c) with $U(T, r, c) \neq \emptyset$, and $D_{IDR} = \{(\alpha, |U(T, r, c)|, \Delta(\alpha, T, r, c)) : \alpha \in \mathbb{R}, 0 \leq \alpha < 1, (T, r, c) \in P_{IDR}\}$.

Lemma 3.37. $D_{IDR} = D_{SC}$.

Proof. Let α be a real number, $0 \leq \alpha < 1$ and $(T, r, c) \in P_{IDR}$. By Proposition 3.27, $(\alpha, |U(T, r, c)|, \Delta(\alpha, T, r, c)) = (\alpha, |A(T, r, c)|, \Delta(\alpha, A(T, r, c), S(T, r, c)))$. Therefore, $D_{IDR} \subseteq D_{SC}$.

Let α be a real number, $0 \leq \alpha < 1$ and $(A, S) \in P_{SC}$. By Proposition 3.29, $(\alpha, |A|, \Delta(\alpha, A, S)) = (\alpha, |U(T(A, S), r(A, S), c(A, S))|, \Delta(\alpha, T(A, S), r(A, S), c(A, S)))$. Therefore, $D_{SC} \subseteq D_{IDR}$. $\qquad \square$

Note that the set D_{SC} was described in Lemma 3.16.

The Best Upper Bound on $L_{\min}(\alpha)$

We define a function $\mathcal{U}_{IDR} : D_{IDR} \to \mathbb{N}$. Let $(\alpha, n, (\delta_1, \ldots, \delta_t)) \in D_{IDR}$. Then $\mathcal{U}_{IDR}(\alpha, n, (\delta_1, \ldots, \delta_t)) = \max\{L_{\min}(\alpha, T, r, c) : (T, r, c) \in P_{IDR}, |U(T, r, c)| = n, \Delta(\alpha, T, r, c) = (\delta_1, \ldots, \delta_t)\}$. It is clear that

$$L_{\min}(\alpha, T, r, c) \leq \mathcal{U}_{IDR}(\alpha, |U(T, r, c)|, \Delta(\alpha, T, r, c))$$

is the best upper bound on $L_{\min}(\alpha)$ depending on α, $|U(T, r, c)|$, $\Delta(\alpha, T, r, c)$.

Theorem 3.38. *Let* $(\alpha, n, (\delta_1, \ldots, \delta_t)) \in D_{IDR}$. *Then*

$$\mathcal{U}_{IDR}(\alpha, n, (\delta_1, \ldots, \delta_t)) = t .$$

Proof. Let (T, r, c) be an arbitrary inhibitory decision rule problem such that $|U(T, r, c)| = n$ and $\Delta(\alpha, T, r, c) = (\delta_1, \ldots, \delta_t)$. It is clear that $L_{\min}(\alpha, T, r, c) \leq L_{\text{greedy}}(\alpha, T, r, c) = t$. Therefore, $\mathcal{U}_{IDR}(\alpha, n, (\delta_1, \ldots, \delta_t)) \leq t$.

Let us show that $\mathcal{U}_{IDR}(\alpha, n, (\delta_1, \ldots, \delta_t)) \geq t$. Using Lemma 3.37 we obtain $(\alpha, n, (\delta_1, \ldots, \delta_t)) \in D_{SC}$. From here and from Theorem 3.17 it follows that there exists a set cover problem (A, S) such that $|A| = n$, $\Delta(\alpha, A, S) = (\delta_1, \ldots, \delta_t)$ and $C_{\min}(\alpha, A, S) = t$. Let us consider the inhibitory decision rule problem $(T, r, c) = (T(A, S), r(A, S), c(A, S))$. From Proposition 3.29 it follows that $|U(T, r, c)| = n$, $\Delta(\alpha, T, r, c) = (\delta_1, \ldots, \delta_t)$ and $L_{\min}(\alpha, T, r, c) = t$. Therefore, $\mathcal{U}_{IDR}(\alpha, n, (\delta_1, \ldots, \delta_t)) \geq t$. $\qquad \square$

Thus, $L_{\min}(\alpha, T, r, c) \leq L_{\text{greedy}}(\alpha, T, r, c)$ is the best upper bound on $L_{\min}(\alpha)$ depending on α, $|U(T, r, c)|$ and $\Delta(\alpha, T, r, c)$.

The Best Lower Bound on $L_{\min}(\alpha)$

We define a function $\mathcal{L}_{IDR} : D_{IDR} \to \mathbb{N}$. Let $(\alpha, n, (\delta_1, \ldots, \delta_t)) \in D_{IDR}$. Then $\mathcal{L}_{IDR}(\alpha, n, (\delta_1, \ldots, \delta_t)) = \min\{L_{\min}(\alpha, T, r, c) : (T, r, c) \in P_{IDR}, |U(T, r, c)| = n, \Delta(\alpha, T, r, c) = (\delta_1, \ldots, \delta_t)\}$. It is clear that

$$L_{\min}(\alpha, T, r, c) \geq \mathcal{L}_{DR}(\alpha, |U(T, r, c)|, \Delta(\alpha, T, r, c))$$

is the best lower bound on $L_{\min}(\alpha)$ depending on α, $|U(T, r, c)|$, $\Delta(\alpha, T, r, c)$.

Let $(\alpha, n, (\delta_1, \ldots, \delta_t)) \in D_{IDR}$. We now remind the definition of parameter $l(\alpha, n, (\delta_1, \ldots, \delta_t))$. Set $\delta_0 = 0$. Then

$$l(\alpha, n, (\delta_1, \ldots, \delta_t)) = \max \left\{ \left\lceil \frac{\lceil (1 - \alpha)n \rceil - (\delta_0 + \ldots + \delta_i)}{\delta_{i+1}} \right\rceil : i = 0, \ldots, t - 1 \right\}.$$

Theorem 3.39. *Let* $(\alpha, n, (\delta_1, \ldots, \delta_t)) \in D_{IDR}$. *Then*

$$\mathcal{L}_{IDR}(\alpha, n, (\delta_1, \ldots, \delta_t)) = l(\alpha, n, (\delta_1, \ldots, \delta_t)) .$$

Proof. Let (T, r, c) be an arbitrary inhibitory decision rule problem such that $|U(T, r, c)| = n$ and $\Delta(\alpha, T, r, c) = (\delta_1, \ldots, \delta_t)$. We consider now the set cover problem $(A, S) = (A(T, r, c), S(T, r, c))$. From Proposition 3.27 it follows that $|A| = n$ and $\Delta(\alpha, A, S) = (\delta_1, \ldots, \delta_t)$. Using Theorem 3.18 we obtain $C_{\min}(\alpha, A, S) \geq l(\alpha, n, (\delta_1, \ldots, \delta_t))$. By Proposition 3.27, $C_{\min}(\alpha, A, S) = L_{\min}(\alpha, T, r, c)$. Therefore, we have $L_{\min}(\alpha, T, r, c) \geq l(\alpha, n, (\delta_1, \ldots, \delta_t))$ and $\mathcal{L}_{IDR}(\alpha, n, (\delta_1, \ldots, \delta_t)) \geq l(\alpha, n, (\delta_1, \ldots, \delta_t))$.

Let us show that $\mathcal{L}_{IDR}(\alpha, n, (\delta_1, \ldots, \delta_t)) \leq l(\alpha, n, (\delta_1, \ldots, \delta_t))$. By Lemma 3.37, $(\alpha, n, (\delta_1, \ldots, \delta_t)) \in D_{SC}$. From here and from Theorem 3.18 it follows that there exists a set cover problem (A, S) such that $|A| = n$, $\Delta(\alpha, A, S) =$

$(\delta_1, \ldots, \delta_t)$ and $C_{\min}(\alpha, A, S) = l(\alpha, n, (\delta_1, \ldots, \delta_t))$. Let us consider the inhibitory decision rule problem $(T, r, c) = (T(A, S), r(A, S), c(A, S))$. From Proposition 3.29 it follows that $|U(T, r, c)| = n$, $\Delta(\alpha, T, r, c) = (\delta_1, \ldots, \delta_t)$ and $L_{\min}(\alpha, T, r, c) = l(\alpha, n, (\delta_1, \ldots, \delta_t))$. Therefore, $\mathcal{L}_{IDR}(\alpha, n, (\delta_1, \ldots, \delta_t)) \leq l(\alpha, n, (\delta_1, \ldots, \delta_t))$. □

Thus, $L_{\min}(\alpha, T, r, c) \geq l(\alpha, |U(T, r)|, \Delta(\alpha, T, r, c))$ is the best lower bound on $L_{\min}(\alpha)$ depending on α, $|U(T, r, c)|$ and $\Delta(\alpha, T, r, c)$.

Properties of the Best Lower Bound on $L_{\min}(\alpha)$

We assume that (T, r, c) is an inhibitory decision rule problem from P_{IDR}, and $\alpha \in \mathbb{R}$, $0 \leq \alpha < 1$. Let

$$l_{IDR}(\alpha) = l_{IDR}(\alpha, T, r, c) = l(\alpha, |U(T, r, c)|, \Delta(\alpha, T, r, c)) .$$

Lemma 3.40. *Let* $\alpha_1, \alpha_2 \in \mathbb{R}$ *and* $0 \leq \alpha_1 < \alpha_2 < 1$. *Then*

$$l_{IDR}(\alpha_1) \geq l_{IDR}(\alpha_2) .$$

Proof. Let $\Delta(\alpha_1, T, r, c) = (\delta_1, \ldots, \delta_{t_1})$ and $\Delta(\alpha_2, T, r, c) = (\delta_1, \ldots, \delta_{t_2})$. It is clear that $t_1 \geq t_2$. Set $\delta_0 = 0$. Let $j \in \{0, \ldots, t_2 - 1\}$ and

$$\left\lceil \frac{\lceil |U(T, r, c)|(1 - \alpha_2)\rceil - (\delta_0 + \ldots + \delta_j)}{\delta_{j+1}} \right\rceil = l_{IDR}(\alpha_2) .$$

It is clear that $l_{IDR}(\alpha_1) \geq \lceil (\lceil |U(T, r, c)|(1 - \alpha_1)\rceil - (\delta_0 + \ldots + \delta_j))/\delta_{j+1} \rceil \geq l_{IDR}(\alpha_2)$. □

Corollary 3.41. $l_{IDR}(0) = \max\{l_{IDR}(\alpha) : 0 \leq \alpha < 1\}$.

The value $l_{IDR}(\alpha)$ can be used for obtaining upper bounds on the length of partial inhibitory decision rules constructed by the greedy algorithm.

Theorem 3.42. *Let* α *and* β *be real numbers such that* $0 < \beta \leq \alpha < 1$. *Then*

$$L_{\text{greedy}}(\alpha) < l_{IDR}(\alpha - \beta) \ln \left(\frac{1 - \alpha + \beta}{\beta} \right) + 1 .$$

Proof. Let us denote $(A, S) = (A(T, r, c), S(T, r, c))$. From Theorem 3.21 it follows that $C_{\text{greedy}}(\alpha, A, S) < l_{SC}(\alpha - \beta, A, S) \ln ((1 - \alpha + \beta)/\beta) + 1$. Using Proposition 3.27 one can show that $l_{IDR}(\alpha - \beta) = l_{IDR}(\alpha - \beta, T, r, c) = l_{SC}(\alpha - \beta, A, S)$. From Proposition 3.27 it follows that

$$L_{\text{greedy}}(\alpha) = L_{\text{greedy}}(\alpha, T, r, c) = C_{\text{greedy}}(\alpha, A, S) .$$

Therefore, the statement of the theorem holds. □

Corollary 3.43. *Let* $\alpha \in \mathbb{R}$, $0 < \alpha < 1$. *Then*

$$L_{\text{greedy}}(\alpha) < l_{IDR}(0) \ln \left(\frac{1}{\alpha} \right) + 1 .$$

If $l_{IDR}(0)$ is a small number, then we have a good upper bound on $L_{\text{greedy}}(\alpha)$. If $l_{IDR}(0)$ is a big number, then we have a big lower bound on $L_{\min}(0)$ and on $L_{\min}(\alpha)$ for some α.

3.2.6   Upper Bound on $L_{\text{greedy}}(\alpha)$

We assume that (T, r, c) is an inhibitory decision rule problem from P_{IDR}. In this subsection, we obtain an upper bound on $L_{\text{greedy}}(\alpha) = L_{\text{greedy}}(\alpha, T, r, c)$, which does not depend on $|U(T, r, c)|$, and show that, in some sense, this bound is unimprovable.

Theorem 3.44. *Let α and β be real numbers such that $0 < \beta \leq \alpha < 1$. Then*

$$L_{\text{greedy}}(\alpha) < L_{\min}(\alpha - \beta) \ln \left(\frac{1 - \alpha + \beta}{\beta} \right) + 1 \,.$$

Proof. By Theorem 3.42, $L_{\text{greedy}}(\alpha) < l_{IDR}(\alpha - \beta) \ln \left((1 - \alpha + \beta)/\beta \right) + 1$, and by Theorem 3.39, $l_{IDR}(\alpha - \beta) \leq L_{\min}(\alpha - \beta)$.   □

Let us show that obtained bound is, in some sense, unimprovable.

Theorem 3.45. *There is no real $\delta < 1$ such that for any inhibitory decision rule problem $(T, r, c) \in P_{IDR}$ and for any real α and β, $0 < \beta \leq \alpha < 0$, the following inequality holds:*

$$L_{\text{greedy}}(\alpha) \leq \delta \left(L_{\min}(\alpha - \beta) \ln \left(\frac{1 - \alpha + \beta}{\beta} \right) + 1 \right) \,.$$

Proof. We assume the contrary: let such δ exist. We now consider an arbitrary set cover problem (A, S) and arbitrary real α and β such that $0 < \beta \leq \alpha < 0$. Set $(T, r, c) = (T(A, S), r(A, S), c(A, S))$. Then

$$L_{\text{greedy}}(\alpha, T, r, c) \leq \delta \left(L_{\min}(\alpha - \beta, T, r, c) \ln \left(\frac{1 - \alpha + \beta}{\beta} \right) + 1 \right) \,.$$

By Proposition 3.29, $L_{\text{greedy}}(\alpha, T, r, c) = C_{\text{greedy}}(\alpha, A, S)$ and $L_{\min}(\alpha - \beta, T, r, c) = C_{\min}(\alpha - \beta, A, S)$. Therefore, there exists real $\delta < 1$ such that for any set cover problem (A, S) and for any real α and β, $0 < \beta \leq \alpha < 0$, the inequality $C_{\text{greedy}}(\alpha, A, S) \leq \delta \left(C_{\min}(\alpha - \beta, A, S) \ln \left((1 - \alpha + \beta)/\beta \right) + 1 \right)$ holds, which contradicts Theorem 3.24.   □

3.2.7   Inhibitory Decision Rules for the Most Part of Binary Decision Tables

In this subsection, inhibitory decision rules for the most part of binary decision tables are discussed. In particular, we obtain a confirmation of the following informal 0.5-hypothesis for inhibitory decision rules: for the most part of decision tables T for each row r, and for each decision c, different from decision attached to r, under the construction of partial inhibitory decision rule, during each step the greedy algorithm chooses an attribute which separates from the row r at least one-half of unseparated rows from $U(T, r, c)$.

Tests and Local Tests for the Most Part of Binary Information Systems

A *binary information system* I is a table with n rows (corresponding to objects) and m columns labeled with attributes $a_1, \ldots, a_m$. This table is filled by numbers from $\{0, 1\}$ (values of attributes). For $j = 1, \ldots, n$, we denote by r_j the j-th row of table I.

A subset $\{a_{i_1}, \ldots, a_{i_k}\}$ of attributes is a *test* for the information system I if these attributes separate any two rows r_j and r_l, where $j, l \in \{1, \ldots, n\}$ and $j \neq l$.

Theorem 3.46. (Moshkov, Piliszczuk, Zielosko [47]) *Let us consider binary information systems with n rows and m columns labeled with attributes $a_1, \ldots, a_m$. Let $m \geq \lceil 2 \log_2 n \rceil + t$, where t is a natural number, and $i_1, \ldots, i_{\lceil 2 \log_2 n \rceil + t}$ be different numbers from $\{1, \ldots, m\}$. Then the fraction of information systems, for which $\{a_{i_1}, \ldots, a_{i_{\lceil 2 \log_2 n \rceil + t}}\}$ is a test, is at least*

$$1 - \frac{1}{2^{t+1}} .$$

Adding an arbitrary decision attribute d to the considered information system I we obtain a decision table T. For $j = 1, \ldots, n$, let $r_j = (b_1^j, \ldots, b_m^j)$ and d_j be the decision attached to r_j. If $\{a_{i_1}, \ldots, a_{i_k}\}$ is a test for the information system I, then for any $j \in \{1, \ldots, n\}$ the rule

$$(a_{i_1} = b_{i_1}^j) \wedge \ldots \wedge (a_{i_k} = b_{i_k}^j) \rightarrow d \neq c$$

is a 0-inhibitory decision rule for (T, r_j, c), where $c \in Dec(T)$ and $c \neq d_j$.

We now fix a set D of decision attributes. From Theorem 3.46 it follows, for example, that for 99% of binary decision tables with n rows, $m \geq \lceil 2 \log_2 n \rceil + 6$ conditional attributes and decision attribute from D for each row and each decision different from decision attached to this row there exists corresponding exact inhibitory decision rule which length is equal to $\lceil 2 \log_2 n \rceil + 6$.

It is possible to improve this bound if we will consider inhibitory decision rules not for all rows, but for one fixed row only.

Let $j \in \{1, \ldots, n\}$. A subset $\{a_{i_1}, \ldots, a_{i_k}\}$ of attributes will be called a *j-th local test* for the information system I if these attributes separate from the row r_j any row r_l, where $l \in \{1, \ldots, n\}$ and $l \neq j$.

Adding an arbitrary decision attribute to the considered information system I we obtain a decision table T. Let $r_j = (b_1, \ldots, b_m)$, b be the decision attached to r_j, $c \in Dec(T)$ and $c \neq b$. If $\{a_{i_1}, \ldots, a_{i_k}\}$ is a j-th local test for the information system I, then

$$(a_{i_1} = b_{i_1}) \wedge \ldots \wedge (a_{i_k} = b_{i_k}) \rightarrow d \neq c$$

is a 0-inhibitory decision rule for (T, r_j, c).

Let us fix a set D of decision attributes. If we prove the existence of good j-th local tests for the most part of binary information systems with n rows and m columns, then it means the existence of good inhibitory decision rules for

j-th row for the most part of binary decision tables with n rows, m conditional attributes and decision attributes from D.

Theorem 3.47. (Moshkov, Piliszczuk, Zielosko [47]) *Let us consider binary information systems with n rows and m columns labeled with attributes $a_1, \ldots, a_m$. Let $m \geq \lceil \log_2 n \rceil + t$, where t is a natural number, $j \in \{1, \ldots, n\}$ and $i_1, \ldots, i_{\lceil \log_2 n \rceil + t}$ be pairwise different numbers from $\{1, \ldots, m\}$. Then the fraction of information systems, for which $\{a_{i_1}, \ldots, a_{i_{\lceil \log_2 n \rceil + t}}\}$ is a j-th local test, is at least*

$$1 - \frac{1}{2^t} .$$

Let us fix a set D of decision attributes and a number $j \in \{1, \ldots, n\}$. From obtained result it follows that for 99% of binary decision tables with n rows, $m \geq \lceil \log_2 n \rceil + 7$ conditional attributes and the decision attribute from D, for j-th row for each decision different from the decision attached to the j-th row there exists corresponding exact inhibitory decision rule which length is equal to $\lceil \log_2 n \rceil + 7$.

Partial Inhibitory Decision Rules Constructed by Greedy Algorithm for the Most Part of Binary Decision Tables

We now study the behavior of greedy algorithm for the most part of binary decision tables, under some assumption on relationships between the number of rows and the number of columns in tables.

Let I be a binary information system with n rows and m columns labeled with attributes $a_1, \ldots, a_m$. For $j = 1, \ldots, n$, we denote by r_j the j-th row of I. The information system I will be called *strongly saturated* if, for any row $r_j = (b_1, \ldots, b_m)$ of I, for any $k \in \{1, \ldots, n-1\}$ and for any k rows with numbers different from j, there exists a column a_i which has at least $k/2$ numbers $\neg b_i$ at the intersection with considered k rows.

First, we evaluate the number of strongly saturated binary information systems. After that, we study the work of greedy algorithm on a decision table obtained from a strongly saturated binary information system by adding a decision attribute. It is clear that the 0.5-hypothesis for inhibitory decision rules holds for every such table.

Theorem 3.48. (Moshkov, Piliszczuk, Zielosko [47]) *Let us consider binary information systems with n rows and $m \geq n + \log_2 n$ columns labeled with attributes $a_1, \ldots, a_m$. Then the fraction of strongly saturated information systems is at least*

$$1 - \frac{1}{2^{m-n-\log_2 n + 1}} .$$

For example, if $m \geq n + \log_2 n + 6$, then at least 99% of binary information systems are strongly saturated.

Let us consider the work of greedy algorithm on an arbitrary decision table T obtained from a strongly saturated binary information system. Let r be an arbitrary row of table T and c be a decision different from the decision attached to r. For $i = 1, 2, \ldots$, after the step number i at most $|U(T, r, c)|/2^i$ rows from $U(T, r, c)$ are unseparated from r. It is not difficult to show that $L_{\text{greedy}}(\alpha) \leq \lceil \log_2(1/\alpha) \rceil$ for any real α, $0 < \alpha < 1$. One can prove that $L_{\text{greedy}}(0) \leq \log_2 |U(T, r, c)| + 1$. It is easy to check that $l_{IDR}(0) \leq 2$.

3.3 Conclusions

The chapter is devoted (mainly) to the analysis of greedy algorithm for construction of partial inhibitory decision rules.

The obtained results show that, under some natural assumptions on the class NP, this algorithm is close to the best polynomial approximate algorithms for the minimization of the length of partial inhibitory decision rules. Based on an information received during greedy algorithm work it is possible to obtain lower and upper bounds on the minimal length of rules. Some theoretical results show that, for the most part of binary decision tables, greedy algorithm constructs simple partial inhibitory decision rules with relatively high accuracy. In particular, these results confirm the 0.5-hypothesis for inhibitory decision rules.

4 Partial Covers and Inhibitory Decision Rules with Weights

In this chapter, we consider the case, where each subset, used for covering, has its own weight, and we should minimize the total weight of subsets in partial cover. The same situation is with partial inhibitory decision rules: each conditional attribute has its own weight, and we should minimize the total weight of attributes occurring in partial inhibitory decision rule. If weights of attributes characterize time complexity of attribute value computation, then we try to minimize total time complexity of computation of attributes from partial inhibitory decision rule. If weights characterize a risk of attribute value computation (as in medical or technical diagnosis), then we try to minimize total risk, etc.

Based on results obtained for partial covers we study partial inhibitory decision rules for decision tables which, in general case, are inconsistent (a decision table is inconsistent if it has equal rows with different decisions). In particular, we show that:

- Under some natural assumptions on the class NP, the greedy algorithm with weights is close to the best polynomial approximate algorithms for minimization of the total weight of attributes occurring in partial inhibitory decision rules.
- Based on an information received during the greedy algorithm runs it is possible to obtain nontrivial lower bounds on the minimal total weight of attributes occurring in partial inhibitory decision rules.
- There exist polynomial time modifications of the standard greedy algorithm which for a part of decision tables can give better results than the standard greedy algorithm.

Similar results can be obtained for partial inhibitory association rules with weights over information systems. To this end, it is enough to fix an arbitrary attribute a_i of the information system as the decision attribute and study inhibitory association rules with the right-hand side of the kind $a_i \neq c$ as inhibitory decision rules over the obtained decision system (decision table).

This chapter is, in some sense, an extension of Chap. 3 to the case of weights which are not equal to 1. However, problems considered in this chapter are more

P. Delimata et al.: Inhibitory Rules in Data Analysis, SCI 163, pp. 63–79.

compound than the ones considered in Chap. 3. Bounds obtained in this chapter are sometimes weaker than the corresponding bounds from Chap. 3. We should also note that even if all weights are equal to 1, then results of the work of greedy algorithms considered in this chapter can be different from the results of the work of greedy algorithms considered in Chap. 3. For example, for the case of rules the number of chosen attributes is the same, but the last attributes can differ.

In this chapter, we use the collection of results on partial covers with weights mentioned or obtained in [45, 48] together with some definitions and examples.

The chapter consists of three sections. In Sect. 4.1, known results for partial covers are recalled. In Sect. 4.2, partial inhibitory decision rules are studied. Section 4.3 contains short conclusions.

4.1 Partial Covers with Weights

This section consists of seven subsections. In Sect. 4.1.1, main notions are considered. In Sect. 4.1.2, some known results are reported. In Sect. 4.1.3, polynomial approximate algorithms for minimization of partial cover weight are discussed. In Sect. 4.1.4, a comparison of the standard greedy algorithm and a greedy algorithm with two thresholds is given. Two modifications of the standard greedy algorithm are considered in Sect. 4.1.5. Section 4.1.6 is devoted to the investigation of the lower bound on the minimal weight of partial cover depending on some information about runs of the greedy algorithm with two thresholds. In Sect. 4.1.7, two bounds on precision of the greedy algorithm with two thresholds are considered that do not depend on the cardinality of covered set.

4.1.1 Main Notions

We repeat here some definitions from Chap. 3 and consider generalizations of other definitions to the case of arbitrary natural weights.

Let $A = \{e_1, \ldots, e_n\}$ be a nonempty finite set. Elements of A are enumerated by numbers $1, \ldots, n$ (in fact, we fix a linear order on A). Let $S = \{B_i\}_{i \in \{1,\ldots,m\}} = \{B_1, \ldots, B_m\}$ be a family of subsets of A such that $B_1 \cup \ldots \cup B_m = A$. We will assume that S can contain equal subsets of A. The pair (A, S) will be called a *set cover problem*. Let w be a *weight function* which corresponds to each $B_i \in S$ a natural number $w(B_i)$. The triple (A, S, w) will be called a *set cover problem with weights*. Note that, in fact, the weight function w is given on the set of indexes $\{1, \ldots, m\}$. But, for simplicity, we are writing $w(B_i)$ instead of $w(i)$.

Let I be a subset of $\{1, \ldots, m\}$. The family $P = \{B_i\}_{i \in I}$ will be called a *subfamily* of S. The number $|P| = |I|$ will be called the *cardinality* of P. Let $P = \{B_i\}_{i \in I}$ and $Q = \{B_i\}_{i \in J}$ be subfamilies of S. The notation $P \subseteq Q$ will mean that $I \subseteq J$. Let us denote $P \cup Q = \{B_i\}_{i \in I \cup J}$, $P \cap Q = \{B_i\}_{i \in I \cap J}$, and $P \setminus Q = \{B_i\}_{i \in I \setminus J}$.

A subfamily $Q = \{B_{i_1}, \ldots, B_{i_t}\}$ of the family S will be called a *partial cover for* (A, S). Let α be a real number such that $0 \le \alpha < 1$. The subfamily Q will be called an *α-cover for* (A, S) if $|B_{i_1} \cup \ldots \cup B_{i_t}| \geq (1 - \alpha)|A|$. For example, 0.01-cover means that we should cover at least 99% of elements from A. Note that a 0-cover is the standard (exact) cover. The number $w(Q) = \sum_{j=1}^{t} w(B_{i_j})$ will be called the *weight* of the partial cover Q. Let us denote by $C_{\min}(\alpha) = C_{\min}(\alpha, A, S, w)$ the minimal weight of α-cover for (A, S).

Let α and γ be real numbers such that $0 \leq \gamma \leq \alpha < 1$. We now describe a *greedy algorithm with two thresholds* α *and* γ (see Algorithm 4.1).

Algorithm 4.1. [45] Greedy algorithm with two thresholds α and γ for partial cover construction

---

Input : Set cover problem with weights (A, S, w), where $S = \{B_1, \ldots, B_m\}$, and real numbers α and γ such that $0 \leq \gamma \leq \alpha < 1$.
Output: α-cover for (A, S).
$Q \longleftarrow \emptyset$;
$D \longleftarrow \emptyset$;
$M \longleftarrow \lceil |A|(1 - \alpha) \rceil$;
$N \longleftarrow \lceil |A|(1 - \gamma) \rceil$;
while $|D| < M$ **do**
> select $B_i \in S$ with minimal index i such that $B_i \setminus D \neq \emptyset$ and the value
>
> $$\frac{w(B_i)}{\min\{|B_i \setminus D|, N - |D|\}}$$
>
> is minimal;
> $Q \longleftarrow Q \cup \{B_i\}$;
> $D \longleftarrow D \cup B_i$;

end
return Q;

Let us denote by $C_{\text{greedy}}^{\gamma}(\alpha) = C_{\text{greedy}}^{\gamma}(\alpha, A, S, w)$ the weight of α-cover constructed by the considered algorithm for the set cover problem with weights (A, S, w).

Note that the greedy algorithm with two thresholds α and $\gamma = \alpha$ (greedy algorithm with equal thresholds) coincides with the greedy algorithm with weights considered in [77].

4.1.2   Some Known Results

In this subsection, we assume that the weight function has values from the set of positive real numbers.

For natural m, we denote $H(m) = 1 + \ldots + 1/m$. It is known that

$$\ln m \leq H(m) \leq \ln m + 1 .$$

Let us consider some results for the case of exact covers, where $\alpha = 0$. In this case $\gamma = 0$. The first results belong to Chvátal.

Theorem 4.1. (Chvátal [11]) *For any set cover problem with weights* (A, S, w) *the following inequality holds:*

$$C^0_{\text{greedy}}(0) \leq C_{\min}(0)H(|A|) .$$

Theorem 4.2. (Chvátal [11]) *For any set cover problem with weights* (A, S, w) *the following inequality holds:*

$$C^0_{\text{greedy}}(0) \leq C_{\min}(0)H\left(\max_{B_i \in S} |B_i|\right) .$$

Chvátal proved in [11] that the bounds from Theorems 4.1 and 4.2 are almost unimprovable.

We now consider some results for the case, where $\alpha \geq 0$ and $\gamma = \alpha$. The first upper bound on $C^\alpha_{\text{greedy}}(\alpha)$ was obtained by Kearns.

Theorem 4.3. (Kearns [25]) *For any set cover problem with weights* (A, S, w) *and any* α, $0 \leq \alpha < 1$, *the following inequality holds:*

$$C^\alpha_{\text{greedy}}(\alpha) \leq C_{\min}(\alpha)(2H(|A|) + 3) .$$

This bound was improved by Slavík.

Theorem 4.4. (Slavík [77]) *For any set cover problem with weights* (A, S, w) *and any* α, $0 \leq \alpha < 1$, *the following inequality holds:*

$$C^\alpha_{\text{greedy}}(\alpha) \leq C_{\min}(\alpha)H\left(\lceil (1 - \alpha)|A| \rceil\right) .$$

Theorem 4.5. (Slavík [77]) *For any set cover problem with weights* (A, S, w) *and any* α, $0 \leq \alpha < 1$, *the following inequality holds:*

$$C^\alpha_{\text{greedy}}(\alpha) \leq C_{\min}(\alpha)H\left(\max_{B_i \in S} |B_i|\right) .$$

Slavík proved in [77] that the bounds from Theorems 4.4 and 4.5 are unimprovable.

4.1.3   Polynomial Approximate Algorithms

In this subsection, we consider three theorems which follow immediately from Theorems 3.13, 3.14 and 3.15.

Let $0 \leq \alpha < 1$. We consider the following problem: for a given set cover problem with weights (A, S, w) it is required to find an α-cover for (A, S) with minimal weight.

Theorem 4.6. (Moshkov, Piliszczuk, Zielosko [45]) *Let* $0 \leq \alpha < 1$. *Then the problem of construction of* α-cover *with minimal weight is* NP-hard.

From this theorem it follows that we should consider polynomial approximate algorithms for minimization of α-cover weight.

Theorem 4.7. (Moshkov, Piliszczuk, Zielosko [45]) *Let $\alpha \in \mathbb{R}$ and $0 \leq \alpha < 1$. If $NP \not\subseteq DTIME(n^{O(\log \log n)})$, then for any ε, $0 < \varepsilon < 1$, there is no polynomial algorithm that for a given set cover problem with weights (A, S, w) constructs an α-cover for (A, S) which weight is at most*

$$(1 - \varepsilon)C_{\min}(\alpha, A, S, w) \ln |A| .$$

Theorem 4.8. (Moshkov, Piliszczuk, Zielosko [45]) *Let α be a real number such that $0 \leq \alpha < 1$. If $P \neq NP$, then there exists $\delta > 0$ such that there is no polynomial algorithm that for a given set cover problem with weights (A, S, w) constructs an α-cover for (A, S) which weight is at most*

$$\delta C_{\min}(\alpha, A, S, w) \ln |A| .$$

From Theorem 4.5 it follows that $C^{\alpha}_{\text{greedy}}(\alpha) \leq C_{\min}(\alpha)(1 + \ln |A|)$. From this inequality and from Theorem 4.7 it follows that, under the assumption $NP \not\subseteq DTIME(n^{O(\log \log n)})$, the greedy algorithm with two thresholds α and $\gamma = \alpha$ (in fact, the greedy algorithm with weights from [77]) is close to the best polynomial approximate algorithms for minimization of partial cover weight. From the considered inequality and from Theorem 4.8 it follows that, under the assumption $P \neq NP$, the greedy algorithm with two thresholds α and $\gamma = \alpha$ is not far from the best polynomial approximate algorithms for minimization of partial cover weight.

However, we can try to improve the results of the work of the greedy algorithm with two thresholds α and $\gamma = \alpha$ for some part of set cover problems with weights.

4.1.4   Comparison of Standard Greedy Algorithm and Greedy Algorithm with Two Thresholds

The following example shows that if for the greedy algorithm with two thresholds α and γ we will use γ such that $\gamma < \alpha$, we can obtain sometimes better results than in the case $\gamma = \alpha$.

Example 4.9. (Moshkov, Piliszczuk, Zielosko [45]) Let us consider a set cover problem (A, S, w) such that $A = \{1, 2, 3, 4, 5, 6\}$, $S = \{B_1, B_2\}$, $B_1 = \{1\}$, $B_2 = \{2, 3, 4, 5, 6\}$, $w(B_1) = 1$ and $w(B_2) = 4$. Set $\alpha = 0.5$. It means that we should cover at least $M = \lceil (1 - \alpha)|A| \rceil = 3$ elements from A. If $\gamma = \alpha = 0.5$, then the result of the run of the greedy algorithm with thresholds α and γ is the 0.5-cover $\{B_1, B_2\}$ which weight is equal to 5. If $\gamma = 0 < \alpha$, then the result of the run of greedy algorithm with thresholds α and γ is the 0.5-cover $\{B_2\}$ which weight is equal to 4.

In this subsection, we consider some results from [45] which show that, under some assumptions on $|A|$ and $|S|$, for the most part of set cover problems (A, S) there exists a weight function w and real numbers α, γ such that $0 \leq \gamma < \alpha < 1$ and $C^{\gamma}_{\text{greedy}}(\alpha, A, S, w) < C^{\alpha}_{\text{greedy}}(\alpha, A, S, w)$. First, we consider criterion of existence of such w, α and γ (see Theorem 4.10).

Let A be a finite nonempty set and $S = \{B_1, \ldots, B_m\}$ be a family of subsets of A. We will say that the family S is 1-*uniform* if there exists a natural number k such that $|B_i| = k$ or $|B_i| = k+1$ for any nonempty subset B_i from S. We will say that S is *strongly* 1-*uniform* if S is 1-uniform and for any subsets $B_{l_1}, \ldots, B_{l_t}$ from S the family $\{B_1 \setminus U, \ldots, B_m \setminus U\}$ is 1-uniform, where $U = B_{l_1} \cup \ldots \cup B_{l_t}$.

Theorem 4.10. (Moshkov, Piliszczuk, Zielosko [45]) *Let* (A, S) *be a set cover problem. Then the following two statements are equivalent:*

1. *The family* S *is not strongly* 1-*uniform.*
2. *There exists a weight function* w *and real numbers* α *and* γ *such that* $0 \leq \gamma < \alpha < 1$ *and* $C^{\gamma}_{\text{greedy}}(\alpha, A, S, w) < C^{\alpha}_{\text{greedy}}(\alpha, A, S, w)$.

In [45] it was shown that, under some assumptions on $|A|$ and $|S|$, the most part of set cover problems (A, S) is not 1-uniform, and, therefore, is not strongly 1-uniform.

There is a one-to-one correspondence between set cover problems and tables filled by numbers from $\{0, 1\}$ and having no rows filled by 0 only. Let $A = \{e_1, \ldots, e_n\}$ and $S = \{B_1, \ldots, B_m\}$. Then the problem (A, S) corresponds to the table with n rows and m columns which for $i = 1, \ldots, n$ and $j = 1, \ldots, m$ has 1 at the intersection of i-th row and j-th column if and only if $e_i \in B_j$.

Theorem 4.11. (Moshkov, Piliszczuk, Zielosko [45]) *Let us consider set cover problems* (A, S) *such that* $A = \{e_1, \ldots, e_n\}$ *and* $S = \{B_1, \ldots, B_m\}$. *Let* $n \geq 4$ *and* $m \geq \log_2 n + 1$. *Then the fraction of set cover problems which are not* 1-*uniform is at least*

$$1 - \frac{9^{m/2+1}}{n^{m/2-1}}.$$

So if n is large enough and $m \geq \log_2 n + 1$, then the most part of set cover problems (A, S) with $|A| = n$ and $|S| = m$ is not 1-uniform.

For example, the fraction of set cover problems (A, S) with $|A| = 81$ and $|S| = 20$, which are not 1-uniform, is at least $1 - 1/9^7 = 1 - 1/4782969$.

4.1.5   Two Modifications of Greedy Algorithm

Results discussed in the previous subsection show that the greedy algorithm with two thresholds is of some interest. In this subsection, we consider two polynomial modifications of greedy algorithm studied in [45] which allow us to use advantages of greedy algorithm with two thresholds.

Let (A, S, w) be a set cover problem with weights and α be a real number such that $0 \leq \alpha < 1$.

1. Of course, it is impossible to consider effectively all γ such that $0 \leq \gamma \leq \alpha$. Instead of this, we can consider all natural N such that $M \leq N \leq |A|$, where $M = \lceil |A|(1 - \alpha) \rceil$ (see Algorithm 4.1). For each $N \in \{M, \ldots, |A|\}$, we apply Algorithm 4.1 with parameters M and N to the set cover problem with weights (A, S, w) and after that choose an α-cover with minimal weight among constructed α-covers.

2. There exists also an another way to construct an α-cover which is not worse than the one obtained under consideration of all N such that $M \leq N \leq |A|$. Let us apply greedy algorithm with thresholds α and $\gamma = \alpha$ (see Algorithm 4.1) to the set cover problem with weights (A, S, w). Let the algorithm choose sequentially subsets $B_{g_1}, \ldots, B_{g_t}$. For each $i \in \{0, \ldots, t-1\}$, we find (if it is possible) a subset B_{l_i} from S with minimal weight $w(B_{l_i})$ such that $|B_{g_1} \cup \ldots \cup B_{g_i} \cup B_{l_i}| \geq M$, and form an α-cover $\{B_{g_1}, \ldots, B_{g_i}, B_{l_i}\}$ (if $i = 0$, then it will be the family $\{B_{l_0}\}$). After that, among constructed α-covers $\{B_{g_1}, \ldots, B_{g_t}\}$, ..., $\{B_{g_1}, \ldots, B_{g_i}, B_{l_i}\}$, ... we choose an α-cover with minimal weight. From Proposition 4.12 it follows that the constructed α-cover is not worse than the one constructed under consideration of all γ, $0 \leq \gamma \leq \alpha$, or (which is the same) all N, $M \leq N \leq |A|$.

Proposition 4.12. (Moshkov, Piliszczuk, Zielosko [45]) *Let (A, S, w) be a set cover problem with weights and α, γ be real numbers such that $0 \leq \gamma < \alpha < 1$. Let the greedy algorithm with two thresholds α and α, which is applied to (A, S, w), choose sequentially subsets $B_{g_1}, \ldots, B_{g_t}$. Let the greedy algorithm with two thresholds α and γ, which is applied to (A, S, w), choose sequentially subsets $B_{l_1}, \ldots, B_{l_k}$. Then either $k = t$ and $(l_1, \ldots, l_k) = (g_1, \ldots, g_t)$ or $k \leq t$, $(l_1, \ldots, l_{k-1}) = (g_1, \ldots, g_{k-1})$ and $l_k \neq g_k$.*

4.1.6 Lower Bound on $C_{\min}(\alpha)$

In this subsection, we fix some information about the work of greedy algorithm with two thresholds and consider the best lower bound on the value $C_{\min}(\alpha)$ depending on this information.

Let (A, S, w) be a set cover problem with weights and α, γ be real numbers such that $0 \leq \gamma \leq \alpha < 1$. We now apply the greedy algorithm with thresholds α and γ to the set cover problem with weights (A, S, w). Let during the construction of α-cover the greedy algorithm choose sequentially subsets $B_{g_1}, \ldots, B_{g_t}$.

We denote $B_{g_0} = \emptyset$ and $\delta_0 = 0$. For $i = 1, \ldots, t$, we denote $\delta_i = |B_{g_i} \setminus (B_{g_0} \cup \ldots \cup B_{g_{i-1}})|$ and $w_i = w(B_{g_i})$.

As information on the greedy algorithm work we will use numbers $M_C = M_C(\alpha, \gamma, A, S, w) = \lceil |A|(1 - \alpha) \rceil$ and $N_C = N_C(\alpha, \gamma, A, S, w) = \lceil |A|(1 - \gamma) \rceil$, and tuples $\Delta_C = \Delta_C(\alpha, \gamma, A, S, w) = (\delta_1, \ldots, \delta_t)$, $W_C = W_C(\alpha, \gamma, A, S, w) = (w_1, \ldots, w_t)$.

For $i = 0, \ldots, t - 1$, we denote

$$\varrho_i = \left\lceil \frac{w_{i+1}(M_C - (\delta_0 + \ldots + \delta_i))}{\min\{\delta_{i+1}, N_C - (\delta_0 + \ldots + \delta_i)\}} \right\rceil .$$

Let us define parameter $\varrho_C(\alpha, \gamma) = \varrho_C(\alpha, \gamma, A, S, w)$ as follows:

$$\varrho_C(\alpha, \gamma) = \max \{\varrho_i : i = 0, \ldots, t - 1\} .$$

In [45] it was proved that $\varrho_C(\alpha, \gamma)$ is the best lower bound on $C_{\min}(\alpha)$ depending on M_C, N_C, Δ_C and W_C. This lower bound is based on a generalization of the following simple reasoning: if we should cover M elements, and the maximal cardinality of a subset from S is δ, then we should use at least $\lceil M/\delta \rceil$ subsets.

Theorem 4.13. (Moshkov, Piliszczuk, Zielosko [45]) *For any set cover problem with weights* (A, S, w) *and any real numbers* α, γ, $0 \leq \gamma \leq \alpha < 1$, *the inequality* $C_{\min}(\alpha, A, S, w) \geq \varrho_C(\alpha, \gamma, A, S, w)$ *holds, and there exists a set cover problem with weights* (A', S', w') *such that*

$$M_C(\alpha, \gamma, A', S', w') = M_C(\alpha, \gamma, A, S, w) \,,$$
$$N_C(\alpha, \gamma, A', S', w') = N_C(\alpha, \gamma, A, S, w) \,,$$
$$\Delta_C(\alpha, \gamma, A', S', w') = \Delta_C(\alpha, \gamma, A, S, w) \,,$$
$$W_C(\alpha, \gamma, A', S', w') = W_C(\alpha, \gamma, A, S, w) \,,$$
$$\varrho_C(\alpha, \gamma, A', S', w') = \varrho_C(\alpha, \gamma, A, S, w) \,,$$
$$C_{\min}(\alpha, A', S', w') = \varrho_C(\alpha, \gamma, A', S', w') \,.$$

Let us consider a property of the parameter $\varrho_C(\alpha, \gamma)$ which is important for practical use of the bound from Theorem 4.13.

Proposition 4.14. (Moshkov, Piliszczuk, Zielosko [45]) *Let* (A, S, w) *be a set cover problem with weights and* α, γ *be real numbers such that* $0 \leq \gamma \leq \alpha < 1$. *Then*

$$\varrho_C(\alpha, \alpha, A, S, w) \geq \varrho_C(\alpha, \gamma, A, S, w) \,.$$

4.1.7   Upper Bounds on $C^{\gamma}_{\text{greedy}}(\alpha)$

In this subsection, we consider some properties of parameter $\varrho_C(\alpha, \gamma)$ and two upper bounds on the value $C^{\gamma}_{\text{greedy}}(\alpha)$ which do not depend directly on cardinality of the set A and cardinalities of subsets B_i from S.

Theorem 4.15. (Moshkov, Piliszczuk, Zielosko [45]) *Let* (A, S, w) *be a set cover problem with weights and* α, γ *be real numbers such that* $0 \leq \gamma < \alpha < 1$. *Then*

$$C^{\gamma}_{\text{greedy}}(\alpha, A, S, w) < \varrho_C(\gamma, \gamma, A, S, w) \left(\ln \left(\frac{1 - \gamma}{\alpha - \gamma} \right) + 1 \right) \,.$$

Corollary 4.16. (Moshkov, Piliszczuk, Zielosko [45]) *Let* ε *be a real number, and* $0 < \varepsilon < 1$. *Then for any* α *such that* $\varepsilon \leq \alpha < 1$ *the following inequalities hold:*

$$\varrho_C(\alpha, \alpha) \leq C_{\min}(\alpha) \leq C^{\alpha - \varepsilon}_{\text{greedy}}(\alpha) < \varrho_C(\alpha - \varepsilon, \alpha - \varepsilon) \left(\ln \frac{1}{\varepsilon} + 1 \right) \,.$$

For example, if $\varepsilon = 0.01$ and $0.01 \leq \alpha < 1$, then $\varrho_C(\alpha, \alpha) \leq C_{\min}(\alpha) \leq C^{\alpha - 0.01}_{\text{greedy}}(\alpha) < 5.61 \varrho_C(\alpha - 0.01, \alpha - 0.01)$, and if $\varepsilon = 0.1$ and $0.1 \leq \alpha < 1$, then $\varrho_C(\alpha, \alpha) \leq C_{\min}(\alpha) \leq C^{\alpha - 0.1}_{\text{greedy}}(\alpha) < 3.31 \varrho_C(\alpha - 0.1, \alpha - 0.1)$.

The obtained results show that the lower bound $C_{\min}(\alpha) \geq \varrho_C(\alpha, \alpha)$ is nontrivial.

Theorem 4.17. (Moshkov, Piliszczuk, Zielosko [45]) *Let (A, S, w) be a set cover problem with weights and α, γ be real numbers such that $0 \leq \gamma < \alpha < 1$. Then*

$$C^\gamma_{\text{greedy}}(\alpha, A, S, w) < C_{\min}(\gamma, A, S, w) \left(\ln \left(\frac{1 - \gamma}{\alpha - \gamma} \right) + 1 \right) .$$

Corollary 4.18. (Moshkov, Piliszczuk, Zielosko [45])

$$C^{0.3}_{\text{greedy}}(0.5) < 2.26 C_{\min}(0.3), C^{0.1}_{\text{greedy}}(0.2) < 3.20 C_{\min}(0.1) ,$$
$$C^{0.001}_{\text{greedy}}(0.01) < 5.71 C_{\min}(0.001), C^0_{\text{greedy}}(0.001) < 7.91 C_{\min}(0) .$$

Corollary 4.19. (Moshkov, Piliszczuk, Zielosko [45]) *Let $0 < \alpha < 1$. Then*

$$C^0_{\text{greedy}}(\alpha) < C_{\min}(0) \left(\ln \left(\frac{1}{\alpha} \right) + 1 \right) .$$

Corollary 4.20. (Moshkov, Piliszczuk, Zielosko [45]) *Let ε be a real number, and $0 < \varepsilon < 1$. Then for any α such that $\varepsilon \leq \alpha < 1$ the following inequalities hold:*

$$C_{\min}(\alpha) \leq C^{\alpha - \varepsilon}_{\text{greedy}}(\alpha) < C_{\min}(\alpha - \varepsilon) \left(\ln \left(\frac{1}{\varepsilon} \right) + 1 \right) .$$

4.2 Partial Inhibitory Decision Rules with Weights

This section consists of six subsections. In Sect. 4.2.1, main notions are considered. In Sect. 4.2.2, some relationships between partial covers and partial inhibitory decision rules are discussed. In Sect. 4.2.3, two bounds on precision of greedy algorithm with thresholds α and $\gamma = \alpha$ are considered. In Sect. 4.2.4, polynomial approximate algorithms for minimization of the weight of partial inhibitory decision rules are studied. Two modifications of greedy algorithm are considered in Sect. 4.2.5. Section 4.2.6 is devoted to consideration of some bounds on minimal weight of partial inhibitory decision rules and weight of inhibitory decision rules constructed by greedy algorithm with thresholds α and γ.

4.2.1 Main Notions

We repeat here some definitions from Chap. 3 and consider generalizations of other definitions to the case of arbitrary natural weights.

Let T be a table with n rows labeled with nonnegative integers (values of the decision attribute d) and m columns labeled with conditional attributes (names of attributes) $a_1, \ldots, a_m$. This table is filled by nonnegative integers (values of conditional attributes). The table T is called a *decision table*. We denote by $Dec(T)$ the set of decisions attached to rows of T. Later we will assume that $|Dec(T)| \geq 2$. Let w be a *weight function* for T which corresponds to each attribute a_i a natural number $w(a_i)$. Let $r = (b_1, \ldots, b_m)$ be a row of T labeled with a decision b, and c be a decision from $Dec(T)$ such that $c \neq b$.

Let us denote by $U(T, r, c)$ the set of rows from T which are different (in at least one column) from r and are labeled with the decision c. We will say that an attribute a_i *separates* rows r and $r' \in U(T, r, c)$ if rows r and r' have different numbers at the intersection with the column a_i. For $i = 1, \ldots, m$, we denote by $U(T, r, c, a_i)$ the set of rows from $U(T, r, c)$ which attribute a_i separates from the row r.

Let α be a real number such that $0 \leq \alpha < 1$. A rule

$$(a_{i_1} = b_{i_1}) \wedge \ldots \wedge (a_{i_t} = b_{i_t}) \to d \neq c \qquad (4.1)$$

is called an α-*inhibitory decision rule* for T, r and c if attributes $a_{i_1}, \ldots, a_{i_t}$ separate from r at least $(1 - \alpha)|U(T, r, c)|$ rows from $U(T, r, c)$. The number $\sum_{j=1}^{t} w(a_{i_j})$ is called the *weight* of the considered decision rule.

If $U(T, r, c) = \emptyset$, then for any $a_{i_1}, \ldots, a_{i_t} \in \{a_1, \ldots, a_m\}$ the rule (4.1) is an α-inhibitory decision rule for T, r and c. Also, the rule (4.1) with empty left-hand side (where $t = 0$) is an α-inhibitory decision rule for T, r and c. The weight of this rule is equal to 0.

For example, 0.01-inhibitory decision rule means that we should separate from r at least 99% of rows from $U(T, r, c)$. Note that 0-rule is the standard (exact) rule. Let us denote by $L_{\min}(\alpha) = L_{\min}(\alpha, T, r, c, w)$ the minimal weight of α-inhibitory decision rule for T, r and c.

Let α, γ be real numbers such that $0 \leq \gamma \leq \alpha < 0$. We now describe a *greedy algorithm with thresholds* α and γ which constructs an α-inhibitory decision rule for given T, r, c and weight function w (see Algorithm 4.2).

Algorithm 4.2. Greedy algorithm with two thresholds α and γ for partial inhibitory decision rule construction

---

Input   : Decision table T with conditional attributes $a_1, \ldots, a_m$, and decision attribute d, row $r = (b_1, \ldots, b_m)$ of T labeled with the decision b, decision $c \in Dec(T)$ such that $c \neq b$, weight function $w : \{a_1, \ldots, a_m\} \to \mathbb{N}$, and real numbers α and γ such that $0 \leq \gamma \leq \alpha < 1$.

Output: α-inhibitory decision rule for T, r, and c.

$Q \longleftarrow \emptyset$;
$D \longleftarrow \emptyset$;
$M \longleftarrow \lceil |U(T, r, c)|(1 - \alpha) \rceil$;
$N \longleftarrow \lceil |U(T, r, c)|(1 - \gamma) \rceil$;
while $|D| < M$ **do**

> select $a_i \in \{a_1, \ldots, a_m\}$ with minimal index i such that $U(T, r, c, a_i) \setminus D \neq \emptyset$ and the value
>
> $$\frac{w(a_i)}{\min\{|U(T, r, c, a_i) \setminus D|, N - |D|\}}$$
>
> is minimal;
> $Q \longleftarrow Q \cup \{a_i\}$;
> $D \longleftarrow D \cup U(T, r, c, a_i)$;

end
return $\bigwedge_{a_i \in Q}(a_i = b_i) \to d \neq c$;

Let us denote by $L_{\text{greedy}}^{\gamma}(\alpha) = L_{\text{greedy}}^{\gamma}(\alpha, T, r, c, w)$ the weight of α-decision rule constructed by the considered algorithm for given table T, row r, decision c and weight function w.

4.2.2 Relationships between Partial Covers and Partial Inhibitory Decision Rules

Let (A, S, w) be a set cover problem with weights and α, γ be real numbers such that $0 \leq \gamma \leq \alpha < 1$. We now apply the greedy algorithm with thresholds α and γ to (A, S, w). Let during the construction of α-cover the greedy algorithm choose sequentially subsets $B_{j_1}, \ldots, B_{j_t}$ from the family S. We denote $O_C(\alpha, \gamma, A, S, w) = (j_1, \ldots, j_t)$.

Let T be a decision table with decision attribute d and m columns labeled with conditional attributes $a_1, \ldots, a_m$, r be a row from T labeled with the decision b, $c \in Dec(T)$, $c \neq b$, and w be a weight function for T. Let $U(T, r, c)$ be a nonempty set.

We correspond a set cover problem with weights $(A(T, r, c), S(T, r, c), u_w)$ to the considered decision table T, row r, decision c and weight function w in the following way:

$$A(T, r, c) = U(T, r, c), \quad S(T, r, c) = \{B_1(T, r, c), \ldots, B_m(T, r, c)\},$$

where $B_1(T, r, c) = U(T, r, c, a_1), \ldots, B_m(T, r, c) = U(T, r, c, a_m)$, and

$$u_w(B_1(T, r, c)) = w(a_1), \ldots, u_w(B_m(T, r, c)) = w(a_m).$$

Let α, γ be real numbers such that $0 \leq \gamma \leq \alpha < 1$. We now apply the greedy algorithm with thresholds α and γ to decision table T, row r, decision c and weight function w. Let during the construction of α-inhibitory decision rule the greedy algorithm choose sequentially attributes $a_{j_1}, \ldots, a_{j_t}$. We denote $O_L(\alpha, \gamma, T, r, c, w) = (j_1, \ldots, j_t)$.

Set $U(T, r, c, a_{j_0}) = \emptyset$. For $i = 1, \ldots, t$, we denote $w_i = w(a_{j_i})$ and

$$\delta_i = |U(T, r, c, a_{j_i}) \setminus (U(T, r, c, a_{j_0}) \cup \ldots \cup U(T, r, c, a_{j_{i-1}}))|.$$

Set

$$M_L(\alpha, \gamma, T, r, c, w) = \lceil |U(T, r, c)|(1 - \alpha) \rceil,$$
$$N_L(\alpha, \gamma, T, r, c, w) = \lceil |U(T, r, c)|(1 - \gamma) \rceil,$$
$$\Delta_L(\alpha, \gamma, T, r, c, w) = (\delta_1, \ldots, \delta_t),$$
$$W_L(\alpha, \gamma, T, r, c, w) = (w_1, \ldots, w_t).$$

It is not difficult to prove the following statement.

Proposition 4.21. *Let T be a decision table with m columns labeled with attributes $a_1, \ldots, a_m$, r be a row of T labeled with the decision b, $c \in Dec(T)$,*

$c \neq b$, $U(T,r,c) \neq \emptyset$, w be a weight function for T, and α, γ be real numbers such that $0 \leq \gamma \leq \alpha < 1$. Then

$$|U(T,r,c)| = |A(T,r,c)|\,,$$
$$|U(T,r,c,a_i)| = |B_i(T,r,c)|,\ \ i = 1,\ldots,m\,,$$
$$O_L(\alpha,\gamma,T,r,c,w) = O_C(\alpha,\gamma,A(T,r,c),S(T,r,c),u_w)\,,$$
$$M_L(\alpha,\gamma,T,r,c,w) = M_C(\alpha,\gamma,A(T,r,c),S(T,r,c),u_w)\,,$$
$$N_L(\alpha,\gamma,T,r,c,w) = N_C(\alpha,\gamma,A(T,r,c),S(T,r,c),u_w)\,,$$
$$\Delta_L(\alpha,\gamma,T,r,c,w) = \Delta_C(\alpha,\gamma,A(T,r,c),S(T,r,c),u_w)\,,$$
$$W_L(\alpha,\gamma,T,r,c,w) = W_C(\alpha,\gamma,A(T,r,c),S(T,r,c),u_w)\,,$$
$$L_{\min}(\alpha,T,r,c,w) = C_{\min}(\alpha,A(T,r,c),S(T,r,c),u_w)\,,$$
$$L^{\gamma}_{\text{greedy}}(\alpha,T,r,c,w) = C^{\gamma}_{\text{greedy}}(\alpha,A(T,r,c),S(T,r,c),u_w)\,.$$

Let (A,S,w) be a set cover problem with weights, where $A = \{e_1,\ldots,e_n\}$ and $S = \{B_1,\ldots,B_m\}$. We correspond a decision table $T(A,S)$, row $r(A,S)$ of $T(A,S)$, decision $c(A,S)$ different from the decision attached to $r(A,S)$ and a weight function v_w for $T(A,S)$ to the set cover problem with weights (A,S,w) in the following way. The table $T(A,S)$ contains m columns labeled with attributes $a_1,\ldots,a_m$ and $n+1$ rows filled by numbers from $\{0,1\}$. For $i = 1,\ldots,n$ and $j = 1,\ldots,m$, at the intersection of i-th row and j-th column the number 1 stays if and only if $e_i \in B_j$. The row number $n+1$ is filled by 0. Let us describe the values of the decision attribute d. The first n rows are labeled with the decision 0. The last row is labeled with the decision 1. We denote by $r(A,S)$ the last row of $T(A,S)$, and $c(A,S) = 0$. Let $v_w(a_1) = w(B_1),\ldots,v_w(a_m) = w(B_m)$.

For $i = 1,\ldots,n+1$, we denote by r_i the i-th row. It is not difficult to see that $U(T(A,S),r(A,S),c(A,S)) = \{r_1,\ldots,r_n\}$. Let $i \in \{1,\ldots,n\}$ and $j \in \{1,\ldots,m\}$. One can show that the attribute a_j separates the row $r_{n+1} = r(A,S)$ from the row r_i if and only if $e_i \in B_j$.

It is not difficult to prove the following statement.

Proposition 4.22. *Let (A,S,w) be a set cover problem with weights and α, γ be real numbers such that $0 \leq \gamma \leq \alpha < 1$. Then*

$$|U(T(A,S),r(A,S),c(A,S))| = |A|\,,$$
$$O_L(\alpha,\gamma,T(A,S),r(A,S),c(A,S),v_w) = O_C(\alpha,\gamma,A,S,w)\,,$$
$$M_L(\alpha,\gamma,T(A,S),r(A,S),c(A,S),v_w) = M_C(\alpha,\gamma,A,S,w)\,,$$
$$N_L(\alpha,\gamma,T(A,S),r(A,S),c(A,S),v_w) = N_C(\alpha,\gamma,A,S,w)\,,$$
$$\Delta_L(\alpha,\gamma,T(A,S),r(A,S),c(A,S),v_w) = \Delta_C(\alpha,\gamma,A,S,w)\,,$$
$$W_L(\alpha,\gamma,T(A,S),r(A,S),c(A,S),v_w) = W_C(\alpha,\gamma,A,S,w)\,,$$
$$L_{\min}(\alpha,T(A,S),r(A,S),c(A,S),v_w) = C_{\min}(\alpha,A,S,w)\,,$$
$$L^{\gamma}_{\text{greedy}}(\alpha,T(A,S),r(A,S),c(A,S),v_w) = C^{\gamma}_{\text{greedy}}(\alpha,A,S,w)\,.$$

4.2.3   Precision of Greedy Algorithm with Equal Thresholds

The following two statements are simple corollaries of results of Slavík (see Theorems 4.4 and 4.5) and Proposition 4.21.

Theorem 4.23. *Let T be a decision table, r be a row of T, c be a decision different from the decision attached to r, $U(T,r,c) \neq \emptyset$, w be a weight function for T, and α be a real number such that $0 \leq \alpha < 1$. Then*

$$L^\alpha_{\text{greedy}}(\alpha) \leq L_{\min}(\alpha) H\left(\lceil (1-\alpha)|U(T,r,c)|\rceil\right) .$$

Theorem 4.24. *Let T be a decision table with m columns labeled with attributes $a_1, \ldots, a_m$, r be a row of T, c be a decision different from the decision attached to r, $U(T,r,c) \neq \emptyset$, w be a weight function for T, $\alpha \in \mathbb{R}$, $0 \leq \alpha < 1$. Then*

$$L^\alpha_{\text{greedy}}(\alpha) \leq L_{\min}(\alpha) H\left(\max_{i \in \{1,\ldots,m\}} |U(T,r,c,a_i)|\right) .$$

4.2.4   Polynomial Approximate Algorithms

In this subsection, we consider three theorems which follow immediately from Theorems 3.34–3.36.

Let $0 \leq \alpha < 1$. We now consider the following problem: for a given decision table T, row r of T, decision c different from the decision attached to r and weight function w for T it is required to find an α-inhibitory decision rule for T, r and c with minimal weight.

Theorem 4.25. *Let $0 \leq \alpha < 1$. Then the problem of construction of α-inhibitory decision rule with minimal weight is NP-hard.*

So we should consider polynomial approximate algorithms for minimization of α-inhibitory decision rule weight.

Theorem 4.26. *Let $\alpha \in \mathbb{R}$ and $0 \leq \alpha < 1$. If $NP \not\subseteq DTIME(n^{O(\log \log n)})$, then for any ε, $0 < \varepsilon < 1$, there is no polynomial algorithm that for a given decision table T, row r of T, decision c different from the decision attached to r and such that $U(T,r,c) \neq \emptyset$, and weight function w for T constructs an α-inhibitory decision rule for T, r and c which weight is at most*

$$(1-\varepsilon)L_{\min}(\alpha, T, r, c, w) \ln |U(T,r,c)| .$$

Theorem 4.27. *Let α be a real number such that $0 \leq \alpha < 1$. If $P \neq NP$, then there exists $\delta > 0$ such that there is no polynomial algorithm that for a given decision table T, row r of T, decision c different from the decision attached to r and such that $U(T,r,c) \neq \emptyset$, and weight function w for T constructs an α-irreducible decision rule for T, r and c which weight is at most*

$$\delta L_{\min}(\alpha, T, r, c, w) \ln |U(T,r,c)| .$$

From Theorem 4.24 it follows that $L^{\alpha}_{\text{greedy}}(\alpha) \leq L_{\min}(\alpha)(1+\ln|U(T,r,c)|)$. From this inequality and from Theorem 4.26 it follows that, under the assumption $NP \nsubseteq DTIME(n^{O(\log \log n)})$, the greedy algorithm with two thresholds α and $\gamma = \alpha$ is close to the best polynomial approximate algorithms for minimization of partial inhibitory decision rule weight. From the considered inequality and from Theorem 4.27 it follows that, under the assumption $P \neq NP$, the greedy algorithm with two thresholds α and $\gamma = \alpha$ is not far from the best polynomial approximate algorithms for minimization of partial inhibitory decision rule weight.

However, we can try to improve the results of the work of greedy algorithm with two thresholds α and $\gamma = \alpha$ for some part of decision tables.

4.2.5   Two Modifications of Greedy Algorithm

The results obtained for partial covers show that the greedy algorithm with two thresholds α and γ is of some interest. We now consider two polynomial modifications of greedy algorithm for partial inhibitory decision rules which allow us to use advantages of the greedy algorithm with two thresholds α and γ.

Let T be a decision table with decision attribute d and m columns labeled with attributes $a_1, \ldots, a_m$, $r = (b_1, \ldots, b_m)$ be a row of T labeled with decision b, $c \in Dec(T)$, $c \neq b$, $U(T,r,c) \neq \emptyset$, w be a weight function for T and α be a real number such that $0 \leq \alpha < 1$.

1. It is impossible to consider effectively all γ such that $0 \leq \gamma \leq \alpha$. Instead of this, we can consider all natural N such that $M \leq N \leq |U(T,r,c)|$, where $M = \lceil |U(T,r,c)|(1-\alpha) \rceil$ (see Algorithm 4.2). For each $N \in \{M, \ldots, |U(T,r,c)|\}$, we apply Algorithm 4.2 with parameters M and N to T, r, c and w and after that choose an α-inhibitory decision rule with minimal weight among constructed α-inhibitory decision rules.

2. There exists also an another way to construct an α-inhibitory decision rule which is not worse than the one obtained under consideration of all N such that $M \leq N \leq |U(T,r,c)|$. We now apply Algorithm 4.2 with thresholds α and $\gamma = \alpha$ to T, r, c and w. Let the algorithm choose sequentially attributes $a_{j_1}, \ldots, a_{j_t}$. For each $i \in \{0, \ldots, t-1\}$, we find (if it is possible) an attribute a_{l_i} of T with minimal weight $w(a_{l_i})$ such that the rule $(a_{j_1} = b_{j_1}) \wedge \ldots \wedge (a_{j_i} = b_{j_i}) \wedge (a_{l_i} = b_{l_i}) \rightarrow d \neq c$ is an α-inhibitory decision rule for T, r and c (if $i = 0$, then it will be the rule $(a_{l_0} = b_{l_0}) \rightarrow d \neq c$). After that, among constructed α-inhibitory decision rules $(a_{j_1} = b_{j_1}) \wedge \ldots \wedge (a_{j_t} = b_{j_t}) \rightarrow d \neq c$, $\ldots$, $(a_{j_1} = b_{j_1}) \wedge \ldots \wedge (a_{j_i} = b_{j_i}) \wedge (a_{l_i} = b_{l_i}) \rightarrow d \neq c$, $\ldots$ we choose an α-inhibitory decision rule with minimal weight. From Proposition 4.28 it follows that the constructed α-inhibitory decision rule is not worse than the one constructed under consideration of all γ, $0 \leq \gamma \leq \alpha$, or (which is the same) all N, $M \leq N \leq |U(T,r,c)|$.

Using Propositions 4.12 and 4.21 one can prove the following statement.

Proposition 4.28. *Let T be a decision table, r be a row of T, c be a decision different from the decision attached to r, $U(T, r, c) \neq \emptyset$, w be a weight function for T and α, γ be real numbers such that $0 \leq \gamma < \alpha < 1$. Let the greedy algorithm with two thresholds α and α, which is applied to T, r, c and w, choose sequentially attributes $a_{g_1}, \dots, a_{g_t}$. Let the greedy algorithm with two thresholds α and γ, which is applied to T, r, c and w, choose sequentially attributes $a_{l_1}, \dots, a_{l_k}$. Then either $k = t$ and $(l_1, \dots, l_k) = (g_1, \dots, g_t)$ or $k \leq t$, $(l_1, \dots, l_{k-1}) = (g_1, \dots, g_{k-1})$ and $l_k \neq g_k$.*

4.2.6 Bounds on $L_{\min}(\alpha)$ and $L_{\text{greedy}}^{\gamma}(\alpha)$

First, we fix some information about the work of greedy algorithm with two thresholds and find the best lower bound on the value $L_{\min}(\alpha)$ depending on this information.

Let T be a decision table, r be a row of T, c be a decision different from the decision attached to r and such that $U(T, r, c) \neq \emptyset$, w be a weight function for T, and α, γ be real numbers such that $0 \leq \gamma \leq \alpha < 1$. We now apply the greedy algorithm with thresholds α and γ to the decision table T, row r, decision c and the weight function w. Let during the construction of α-inhibitory decision rule the greedy algorithm choose sequentially attributes $a_{g_1}, \dots, a_{g_t}$.

Let us denote $U(T, r, c, a_{g_0}) = \emptyset$ and $\delta_0 = 0$. For $i = 1, \dots, t$, we denote $\delta_i = |U(T, r, c, a_{g_i}) \setminus (U(T, r, c, a_{g_0}) \cup \dots \cup U(T, r, c, a_{g_{i-1}}))|$ and $w_i = w(a_{g_i})$. As information on the greedy algorithm work we will use numbers

$$M_L = M_L(\alpha, \gamma, T, r, c, w) = \lceil |U(T, r, c)|(1 - \alpha) \rceil \ ,$$
$$N_L = N_L(\alpha, \gamma, T, r, c, w) = \lceil |U(T, r, c)|(1 - \gamma) \rceil$$

and tuples

$$\Delta_L = \Delta_L(\alpha, \gamma, T, r, c, w) = (\delta_1, \dots, \delta_t) \ ,$$
$$W_L = W_L(\alpha, \gamma, T, r, c, w) = (w_1, \dots, w_t) \ .$$

For $i = 0, \dots, t - 1$, we denote

$$\varrho_i = \left\lceil \frac{w_{i+1}(M_L - (\delta_0 + \dots + \delta_i))}{\min\{\delta_{i+1}, N_L - (\delta_0 + \dots + \delta_i)\}} \right\rceil \ .$$

Let us define parameter $\varrho_L(\alpha, \gamma) = \varrho_L(\alpha, \gamma, T, r, c, w)$ as follows:

$$\varrho_L(\alpha, \gamma) = \max\{\varrho_i : i = 0, \dots, t - 1\} \ .$$

We will show that $\varrho_L(\alpha, \gamma)$ is the best lower bound on $L_{\min}(\alpha)$ depending on M_L, N_L, Δ_L and W_L. Next statement follows from Theorem 4.13 and Propositions 4.21 and 4.22.

Theorem 4.29. *For any decision table T, any row r of T, any decision c different from the decision attached to r such that $U(T, r, c) \neq \emptyset$, any weight*

function w for T, and any real numbers α, γ, $0 \leq \gamma \leq \alpha < 1$, the inequality
$L_{\min}(\alpha, T, r, c, w) \geq \varrho_L(\alpha, \gamma, T, r, c, w)$ *holds, and there exists a decision table*
T', *a row* r' *of* T', *a decision* c' *different from the decision attached to* r' *and a*
weight function w' *for* T' *such that*

$$M_L(\alpha, \gamma, T', r', c', w') = M_L(\alpha, \gamma, T, r, c, w) \, ,$$
$$N_L(\alpha, \gamma, T', r', c', w') = N_L(\alpha, \gamma, T, r, c, w) \, ,$$
$$\Delta_L(\alpha, \gamma, T', r', c', w') = \Delta_L(\alpha, \gamma, T, r, c, w) \, ,$$
$$W_L(\alpha, \gamma, T', r', c', w') = W_L(\alpha, \gamma, T, r, c, w) \, ,$$
$$\varrho_L(\alpha, \gamma, T', r', c', w') = \varrho_L(\alpha, \gamma, T, r, c, w) \, ,$$
$$L_{\min}(\alpha, T', r', c', w') = \varrho_L(\alpha, \gamma, T', r', c', w') \, .$$

Let us consider a property of the parameter $\varrho_L(\alpha, \gamma)$ which is important for practical use of the bound from Theorem 4.29. Next statement follows from Propositions 4.14 and 4.21.

Proposition 4.30. *Let* T *be a decision table,* r *be a row of* T, c *be a decision different from the decision attached to* r, $U(T, r, c) \neq \emptyset$, w *be a weight function for* T, *and* α, γ *be real numbers such that* $0 \leq \gamma \leq \alpha < 1$. *Then*

$$\varrho_L(\alpha, \alpha, T, r, c, w) \geq \varrho_L(\alpha, \gamma, T, r, c, w) \, .$$

We now study some properties of parameter $\varrho_L(\alpha, \gamma)$ and obtain two upper bounds on the value $L_{\text{greedy}}^{\gamma}(\alpha)$ which do not depend directly on cardinality of the set $U(T, r, c)$ and cardinalities of subsets $U(T, r, c, a_i)$.

Next statement follows from Theorem 4.15 and Proposition 4.21.

Theorem 4.31. *Let* T *be a decision table,* r *be a row of* T, c *be a decision different from the decision attached to* r, $U(T, r, c) \neq \emptyset$, w *be a weight function for* T, $\alpha, \gamma \in \mathbb{R}$ *and* $0 \leq \gamma < \alpha < 1$. *Then*

$$L_{\text{greedy}}^{\gamma}(\alpha, T, r, c, w) < \varrho_L(\gamma, \gamma, T, r, c, w) \left(\ln \left(\frac{1-\gamma}{\alpha - \gamma} \right) + 1 \right) \, .$$

Corollary 4.32. *Let* $\varepsilon \in \mathbb{R}$ *and* $0 < \varepsilon < 1$. *Then for any* α, $\varepsilon \leq \alpha < 1$, *the following inequalities hold:*

$$\varrho_L(\alpha, \alpha) \leq L_{\min}(\alpha) \leq L_{\text{greedy}}^{\alpha - \varepsilon}(\alpha) < \varrho_L(\alpha - \varepsilon, \alpha - \varepsilon) \left(\ln \left(\frac{1}{\varepsilon} \right) + 1 \right) \, .$$

For example, $\ln(1/0.01) + 1 < 5.61$ and $\ln(1/0.1) + 1 < 3.31$. The obtained results show that the lower bound $L_{\min}(\alpha) \geq \varrho_L(\alpha, \alpha)$ is nontrivial.

Next statement follows from Theorem 4.17 and Proposition 4.21.

Theorem 4.33. *Let* T *be a decision table,* r *be a row of* T, c *be a decision different from the decision attached to* r, $U(T, r, c) \neq \emptyset$, w *be a weight function for* T, $\alpha, \gamma \in \mathbb{R}$ *and* $0 \leq \gamma < \alpha < 1$. *Then*

$$L_{\text{greedy}}^{\gamma}(\alpha, T, r, c, w) < L_{\min}(\gamma, T, r, c, w) \left(\ln \left(\frac{1-\gamma}{\alpha - \gamma} \right) + 1 \right) \, .$$

Corollary 4.34. $L^{0.3}_{\text{greedy}}(0.5) < 2.26 L_{\min}(0.3)$, $L^{0.1}_{\text{greedy}}(0.2) < 3.20 L_{\min}(0.1)$, $L^{0.001}_{\text{greedy}}(0.01) < 5.71 L_{\min}(0.001)$, $L^0_{\text{greedy}}(0.001) < 7.91 L_{\min}(0)$.

Corollary 4.35. *Let* $0 < \alpha < 1$. *Then*

$$L^0_{\text{greedy}}(\alpha) < L_{\min}(0) \left(\ln\left(\frac{1}{\alpha}\right) + 1 \right).$$

Corollary 4.36. *Let* ε *be a real number, and* $0 < \varepsilon < 1$. *Then for any* α *such that* $\varepsilon \le \alpha < 1$ *the following inequalities hold:*

$$L_{\min}(\alpha) \le L^{\alpha - \varepsilon}_{\text{greedy}}(\alpha) < L_{\min}(\alpha - \varepsilon) \left(\ln\left(\frac{1}{\varepsilon}\right) + 1 \right).$$

4.3   Conclusions

The chapter is devoted (mainly) to the analysis of greedy algorithm with weights for partial inhibitory decision rule construction.

Under some natural assumptions on the class NP, this algorithm is close to the best polynomial approximate algorithms for the minimization of the weight of partial inhibitory decision rules.

The obtained results show that the lower bound on minimal weight of partial inhibitory decision rules, based on an information about the greedy algorithm run, is nontrivial and can be used in practice.

Based on the greedy algorithm with two thresholds we create new polynomial approximate algorithms for minimization of the weight of partial inhibitory decision rules.

5 Classifiers Based on Deterministic and Inhibitory Decision Rules

In this chapter, we consider the following problem of classification (prediction): for a decision table T and a new object v, given by values of conditional attributes from T, it is required to generate a decision corresponding to v.

We compare qualities of classifiers based on exact deterministic and inhibitory decision rules.

The first type of classifiers is the following: for a given decision table we construct for each row an exact deterministic decision rule using the greedy algorithm. The obtained system of rules jointly with simple procedure of voting can be considered as a classifier. A deterministic rule, which is realizable for given object, is a vote "pro" the decision from the right-hand side of the rule.

The second type of classifiers is the following: for a given decision table we construct for each row and each decision, which is different from the decision attached to this row, an exact inhibitory decision rule using the greedy algorithm. The obtained system of rules jointly with simple procedure of voting can be considered as a classifier. An inhibitory rule, which is realizable for given object, is a vote "contra" the decision from the right-hand side of the rule.

This chapter consists of five sections. In Sect. 5.1, we recall the notion of decision table (decision system). In Sect. 5.2, we present the definition of a classifier based on exact deterministic decision rules. Section 5.3 contains the definition of a classifier based on exact inhibitory decision rules. In Sect. 5.4, results of experiments are discussed. Section 5.5 includes short conclusions.

5.1 Decision Tables

Let $T = (U, A, d)$ be a *decision table (decision system)*, where $U = \{u_1, \ldots, u_n\}$ is a finite nonempty set of *objects*, $A = \{a_1, \ldots, a_m\}$ is a finite nonempty set of *conditional attributes* (functions defined on U), and d is the *decision attribute* (function defined on U)[1]. We assume that for each $u_i \in U$ and each $a_j \in A$ the value $a_j(u_i)$ and the value $d(u_i)$ belong to ω, where $\omega = \{0, 1, 2, \ldots\}$ is the set of nonnegative integers.

---

[1] Decision tables are called in machine learning as training samples [33].

P. Delimata et al.: Inhibitory Rules in Data Analysis, SCI 163, pp. 81–86.
springerlink.com   © Springer-Verlag Berlin Heidelberg 2009

We also assume that the decision table $T = (U, A, d)$ is given by a *tabular representation*, i.e., a table with m columns and n rows. Columns of the table are labeled with attributes $a_1, \ldots, a_m$. At the intersection of i-th row and j-th column the value $a_j(u_i)$ is included. For $i = 1, \ldots, n$ the i-th row is labeled with the value $d(u_i)$ of the decision attribute d on the object u_i. For $i = 1, \ldots, n$ we identify the object $u_i \in U$ with the tuple $(a_1(u_i), \ldots, a_m(u_i))$, i.e., the i-th row of the tabular representation of the decision table T. By $Dec(T)$ we denote the set of values of the decision attribute d on objects from U. Later we will assume that $|Dec(T)| \geq 2$.

The set $\mathcal{U}(T) = \omega^m$ is called the *universe* for the decision table T. Besides objects from U we consider also objects from $\mathcal{U}(T) \setminus U$. For any object (tuple) $v \in \mathcal{U}(T)$ and any attribute $a_j \in A$ the value $a_j(v)$ is equal to j-th integer in v.

5.2   Classifiers Based on Deterministic Decision Rules

Let T be a decision table with n rows, decision attribute d and m conditional attributes $a_1, \ldots, a_m$. Let $r = (b_1, \ldots, b_m)$ be a row of T labeled with a decision b. By $U(T, r)$ we denote the set of rows from T which are different (in at least one column) from r and are labeled with decisions different from b. We will say that an attribute a_i *separates* a row $r' \in U(T, r)$ from the row r if the rows r and r' have different numbers at the intersection with column a_i. The pair (T, r) will be called a *deterministic decision rule problem*.

A rule

$$(a_{i_1} = b_{i_1}) \wedge \ldots \wedge (a_{i_t} = b_{i_t}) \rightarrow d = b \tag{5.1}$$

is called an *exact deterministic decision rule* for (T, r) if attributes $a_{i_1}, \ldots, a_{i_t}$ separate from r all rows from $U(T, r)$. If $U(T, r) = \emptyset$, then for any $a_{i_1}, \ldots, a_{i_t} \in \{a_1, \ldots, a_m\}$ the rule (5.1) is an exact deterministic decision rule for (T, r). The rule (5.1) with empty left-hand side (when $t = 0$) is also an exact deterministic decision rule for (T, r).

We now describe a greedy algorithm which constructs an exact deterministic decision rule for (T, r) (see Algorithm 5.1). This algorithm is a special case of an algorithm from [47] for construction of partial deterministic decision rules.

Let us consider classifiers based on exact deterministic decision rules.

For every row r of the decision table T we construct an exact deterministic decision rule for (T, r) by Algorithm 5.1. We denote the obtained family of rules by $Drul(T)$. Note that the same rules can be generated for different r.

The family $Drul(T)$ defines a classifier which for a given new object $v \in \mathcal{U}(T)$ creates a decision for this object using only values of conditional attributes for v. For each decision $b \in Dec(T)$, we compute the number $D_b(v)$ of rules from $Drul(T)$ such that (i) the left-hand side of the considered rule is true for v, and (ii) the right-hand side of the rule is equal to $d = b$. If $D_b(v) > 0$ for at least one decision b, then we choose a decision b for which $D_b(v)$ has the maximal value. Otherwise, we choose some fixed decision b_0.

Algorithm 5.1. Greedy algorithm for exact deterministic decision rule construction

---

Input : Decision table T with decision attribute d, and conditional attributes $a_1, \ldots, a_m$, and row $r = (b_1, \ldots, b_m)$ of T labeled with the decision b.

Output: Exact deterministic decision rule for (T, r).

$Q \longleftarrow \emptyset$;

while *attributes from Q separate from r less than $|U(T, r)|$ rows from $U(T, r)$* **do**

    select $a_i \in \{a_1, \ldots, a_m\}$ with minimal index i such that a_i separates from r the maximal number of rows from $U(T, r)$ unseparated by attributes from Q;

    $Q \longleftarrow Q \cup \{a_i\}$;

end

return $\bigwedge_{a_i \in Q} (a_i = b_i) \rightarrow d = b$;

5.3 Classifiers Based on Inhibitory Decision Rules

Let T be a decision table with n rows, decision attribute d and m conditional attributes $a_1, \ldots, a_m$. Let $r = (b_1, \ldots, b_m)$ be a row of T labeled with a decision b. Let $c \in Dec(T)$ and $c \neq b$. By $U(T, r, c)$ we denote the set of rows from T which are different (in at least one column) from r and are labeled with the decision c. We will say that an attribute a_i *separates* a row $r' \in U(T, r, c)$ from the row r if the rows r and r' have different numbers at the intersection with column a_i. The triple (T, r, c) will be called an *inhibitory decision rule problem*.

A rule
$$(a_{i_1} = b_{i_1}) \wedge \ldots \wedge (a_{i_t} = b_{i_t}) \rightarrow d \neq c \tag{5.2}$$

is called an *exact inhibitory decision rule* for (T, r, c) if attributes $a_{i_1}, \ldots, a_{i_t}$ separate from r all rows from $U(T, r, c)$. If $U(T, r, c) = \emptyset$, then for any $a_{i_1}, \ldots, a_{i_t} \in \{a_1, \ldots, a_m\}$ the rule (5.2) is an exact inhibitory decision rule for (T, r). The rule (5.2) with empty left-hand side (when $t = 0$) is also an exact inhibitory decision rule for (T, r, c).

We now describe a greedy algorithm which constructs an exact inhibitory decision rule for (T, r, c) (see Algorithm 5.2). This algorithm is a special case of Algorithm 3.2 for construction of partial inhibitory decision rules.

Let us consider classifiers based on exact inhibitory decision rules.

For every row r of the decision table T and every decision $c \in Dec(T)$, which is different from the decision attached to r, we construct an exact inhibitory decision rule for (T, r, c) by Algorithm 5.2. We denote the obtained family of rules by $Irul(T)$. Note that the same rules can be generated for different r.

The family $Irul(T)$ defines a classifier which for a given new object $v \in \mathcal{U}(T)$ creates a decision for this object using only values of conditional attributes for v. For each decision $b \in Dec(T)$, we compute the number $I_b(v)$ of rules from $Irul(T)$ such that (i) the left-hand side of the considered rule is true for v, and (ii) the right-hand side of the rule is equal to $d \neq b$. If $I_b(v) > 0$ for at least one decision b, then we choose a decision b for which $I_b(v)$ has the minimal value. Otherwise, we choose some fixed decision b_0.

Algorithm 5.2. Greedy algorithm for exact inhibitory decision rule construction

---

Input : Decision table T with decision attribute d, and conditional attributes $a_1, \ldots, a_m$, row $r = (b_1, \ldots, b_m)$ of T labeled with the decision b, and decision $c \in Dec(T)$, $c \neq b$.

Output: Exact inhibitory decision rule for (T, r, c).

$Q \longleftarrow \emptyset$;

while *attributes from Q separate from r less than $|U(T, r, c)|$ rows from $U(T, r, c)$*

do

> select $a_i \in \{a_1, \ldots, a_m\}$ with minimal index i such that a_i separates from r the maximal number of rows from $U(T, r, c)$ unseparated by attributes from Q;
>
> $Q \longleftarrow Q \cup \{a_i\}$;

end

return $\bigwedge_{a_i \in Q}(a_i = b_i) \rightarrow d \neq c$;

5.4 Results of Experiments

To evaluate the accuracy of classifiers, we can use either train-and-test method or k-fold-cross-validation method. In the first case, we split the initial decision table into training and testing tables, construct a classifier using training table, and apply this classifier to rows from the testing table as to new objects. The *accuracy of classification* is the number of rows (objects) from the testing table, which are properly classified, divided by the number of rows in the testing table. In the second case, we split the initial decision table into k sub-tables, and k times apply train-and-test method using each of k sub-tables as the testing table (the other $k - 1$ sub-tables form the training table). As a result, we obtain k accuracies of classification. The mean of these accuracies is considered as the "final" accuracy of classification. The *error rate of classification* is equal to 1 minus the accuracy of classification.

We have performed experiments with classifiers based on deterministic decision rules (D-classifiers) and classifiers based on inhibitory decision rules (I-classifiers). To evaluate error rate of an algorithm on a decision table we use either train-and-test method or cross-validation method.

The following decision tables from [54] were used in our experiments: monk1 (7 attributes, 124 objects in training set, 432 objects in testing set), monk2 (7 attributes, 169 objects in training set, 432 objects in testing set), monk3 (7 attributes, 122 objects in training set, 432 objects in testing set), lymphography (19 attributes, 148 objects, 10-fold cross-validation), diabetes (9 attributes, 768 objects, 12-fold cross-validation), breast-cancer (10 attributes, 286 objects, 10-fold cross-validation), primary-tumor (18 attributes, 339 objects, 10-fold cross-validation), balance-scale (5 attributes, 625 objects, 10-fold cross-validation), lenses (5 attributes, 24 objects, 10-fold cross-validation), soybean-small (35 attributes, 47 objects, 10-fold cross-validation), soybean-large (35 attributes, 307 objects in training set, 376 objects in testing set), zoo (17 attributes, 101

Table 5.1. Results of experiments on original decision tables

Decision table	D-class.	I-class.	Decision table	D-class.	I-class.
monk1	0.051	0.051	lenses	0.117	0.083
monk2	0.127	0.127	soybean-small	0.395	0.350
monk3	0.051	0.051	soybean-large	0.199	0.207
lymphography	0.107	0.050	zoo	0.048	0.029
diabetes	0.135	0.135	post-operative	0.200	0.156
breast-cancer	0.143	0.143	hayes-roth	0.036	0.000
primary-tumor	0.467	0.176	lung-cancer	0.317	0.258
balance-scale	0.176	0.109	solar-flare	0.022	0.022

Table 5.2. Results of experiments on modified decision tables

Decision table	New tables	D opt	I opt	D average	I average
lymphography	9	1	8	0.278	0.148
primary-tumor	3	0	3	0.174	0.124
balance-scale	4	0	4	0.703	0.556
soybean-large	5	4	0	0.189	0.207
zoo	1	0	1	0.109	0.068
post-operative	8	0	5	0.194	0.158
hayes-roth	4	0	4	0.330	0.214
lung-cancer	7	1	5	0.227	0.150
solar-flare	7	0	7	0.098	0.055

objects, 10-fold cross-validation), post-operative (9 attributes, 90 objects, 10-fold cross-validation), hayes-roth (5 attributes, 132 objects in training set, 28 objects in testing set), lung-cancer (57 attributes, 32 objects, 10-fold cross-validation), solar-flare (13 attributes, 1066 objects in training set, 323 objects in testing set). Continuous attributes are discretized by an algorithm from RSES2 [69]. Missing values in decision tables are filled by an algorithm from RSES2 [69].

Table 5.1 contains results of experiments (error rates) for D-classifiers and I-classifiers on original decision tables from [54].

For 1 decision table the error rate of D-classifier is less than the error rate of I-classifier, for 9 decision tables the error rate of I-classifier is less than the error rate of D-classifier, and for 6 decision tables D-classifier and I-classifier have the same error rate.

Table 5.2 contains results of experiments for D-classifiers and I-classifiers on modified decision tables from [54]. For each initial table from [54] we choose a number of many-valued (with at least three values) attributes different from the decision attribute, and consider each such attribute as new decision attribute. As a result, we obtain the same number of new decision tables as the number of chosen attributes (this number can be found in the column "New tables"). The column "D opt" contains the number of new tables for which the error rate of D-classifier is less than the error rate of I-classifier. The column "I opt" contains the number of new tables for which the error rate of I-classifier is less than the error rate of D-classifier. The column "D average" contains the average error rate of D-classifiers for new tables. The column "I average" contains the average error rate of I-classifiers for new tables.

For 6 new decision tables the error rate of D-classifier is less than the error rate of I-classifier, for 37 new decision tables the error rate of I-classifier is less than the error rate of D-classifier, and for 5 new decision tables D-classifier and I-classifier have the same error rate.

In experiments the DMES system [13] was used.

5.5   Conclusions

In this chapter, we have studied an example of classifiers based on exact deterministic and inhibitory decision rules.

In 46 experiments the error rate of the classifier based on inhibitory rules is less than the error rate of the classifier based on deterministic rules. In 7 experiments the error rate of the classifier based on inhibitory rules is greater than the error rate of the classifier based on deterministic rules. In 11 experiments the classifier based on inhibitory rules and the classifier based on deterministic rules have the same error rates.

There are many other approaches for construction of classifiers based on deterministic decision rules, e.g., it is possible to generalize them to the case of inhibitory rules, but one can not claim that there exists an approach or a small group of approaches which is significantly better than the others.

In Chaps. 6 and 7 we study another, in some sense, a more "objective" approach. We do not construct a subset of the set of all true and realizable rules (this is a "subjective" step). We do not construct the whole set of minimal true and realizable rules (this step is too complicated). Instead of this, using polynomial-time algorithms, we obtain an information about the set of all true and realizable rules. This information is used for the generation of the decision for a given new object.

6 Lazy Classification Algorithms Based on Deterministic and Inhibitory Association Rules

In this chapter, we consider the same classification problem as in Chap. 5: for a given decision table T and a new object v it is required to generate a value of the decision attribute on v using values of conditional attributes on v.

To this end, we divide the decision table T into a number of information systems S_i, $i \in Dec(T)$, where $Dec(T)$ is the set of values of the decision attribute in T. For $i \in Dec(T)$, the information system S_i contains only objects (rows) of T with the value of the decision attribute equal to i.

For each information system S_i, we consider both *deterministic* and *inhibitory* association rules of the following form:

$$(a_1(x) = b_1) \wedge \ldots \wedge (a_t(x) = b_t) \rightarrow a_{t+1}(x) = b_{t+1} \,,$$
$$(a_1(x) = b_1) \wedge \ldots \wedge (a_t(x) = b_t) \rightarrow a_{t+1}(x) \neq b_{t+1}$$

respectively, where $a_1, \ldots, a_{t+1}$ are attributes from S_i and $b_1, \ldots, b_{t+1}$ are values of these attributes. Later we will omit the word "association".

For each information system S_i and a given object v, it is constructed (using polynomial-time algorithm) the so called characteristic table. For any object u from S_i and for any attribute a from S_i, the characteristic table contains the entry encoding information if there exists a rule which (i) is true for each object from S_i; (ii) is realizable for u (the left-hand side of the rule is true for u), (iii) is not true for v, and (iv) has the attribute a on the right-hand side. Based on the characteristic table the decision on the "degree" to which v belongs to S_i is made for any i (to this end we can use different evaluation functions), and a decision i with the maximal "degree" is selected.

We consider two families of classification algorithms: in the first family we use the deterministic rules for construction of characteristic tables, and in the second one we use the inhibitory rules.

Results of experiments show that the algorithms based on inhibitory rules are often better than the algorithms based on deterministic rules.

In the literature, one can find a number of papers which are based on the analogous ideas: instead of construction of huge sets of rules it is possible to extract some information on such sets using algorithms having polynomial time

P. Delimata et al.: Inhibitory Rules in Data Analysis, SCI 163, pp. 87–97.
springerlink.com   © Springer-Verlag Berlin Heidelberg 2009

complexity. Such algorithms can be considered as a kind of lazy learning algorithms [1]. Lazy learning algorithms based on deterministic and inhibitory decision rules are considered in Chap. 7.

Note that the idea to use association rules in decision classes for classification was proposed by A. Skowron and Z. Suraj in [74].

This chapter is based on papers [14, 16].

The chapter consists of four sections. In Sect. 6.1, the notions of deterministic and inhibitory characteristic tables, and the notion of evaluation function are introduced. Section 6.2 contains definitions of two families of lazy classification algorithms. In Sect. 6.3, results of experiments are discussed. Section 6.4 contains short conclusions.

6.1   Characteristic Tables

In this section, we consider some notions which will be used later for description of classification algorithms.

6.1.1   Information Systems

Let $S = (U, A)$ be an *information system*, where $U = \{u_1, \ldots, u_n\}$ is a finite nonempty set of *objects* and $A = \{a_1, \ldots, a_m\}$ is a finite nonempty set of *attributes* (functions defined on U). We assume that for each $u_i \in U$ and each $a_j \in A$ the value $a_j(u_i)$ belongs to ω, where $\omega = \{0, 1, 2, \ldots\}$ is the set of nonnegative integers.

We also assume that the information system $S = (U, A)$ is given by a *tabular representation*, i.e., a table with m columns and n rows. Columns of the table are labeled with attributes $a_1, \ldots, a_m$. At the intersection of i-th row and j-th column the value $a_j(u_i)$ is included. For $i = 1, \ldots, n$, we identify any object $u_i \in U$ with the tuple $(a_1(u_i), \ldots, a_m(u_i))$, i.e., the i-th row of the tabular representation of the information system S.

The set $\mathcal{U}(S) = \omega^m$ is called the *universe* for the information system S. Besides objects from U we consider also objects from $\mathcal{U}(S) \setminus U$. For any object (tuple) $v \in \mathcal{U}(S)$ and any attribute $a_j \in A$ the value $a_j(v)$ is equal to j-th integer in v.

6.1.2   Deterministic Characteristic Tables

Let us consider a rule

$$(a_{j_1}(x) = b_1) \wedge \ldots \wedge (a_{j_t}(x) = b_t) \rightarrow a_k(x) = b_k , \tag{6.1}$$

where $t \geq 0$, $a_{j_1}, \ldots, a_{j_t}, a_k \in A$, $b_1, , \ldots, b_t, b_k \in \omega$, and numbers $j_1, \ldots, j_t, k$ are pairwise different. Such rules are called *deterministic associatiation* rules. Later we will omit the word "association". The rule (6.1) will be called *realizable for an object* $u \in \mathcal{U}(S)$ if $a_{j_1}(u) = b_1, \ldots, a_{j_t}(u) = b_t$. The rule (6.1) will be called *true for an object* $u \in \mathcal{U}(S)$ if $a_k(u) = b_k$ or (6.1) is not realizable for u. The rule (6.1) will be called *true for* S if it is true for any object from U. The rule (6.1) will be called *realizable for* S if it is realizable for at least one object from U. We denote by $Det_A(S)$ the set of all deterministic rules each of which is true for S and realizable for S.

Let $u_i \in U, v \in \mathcal{U}(S), a_k \in A$ and $a_k(u_i) \neq a_k(v)$. We say that a rule (6.1) from $Det_A(S)$ *contradicts* v *relative to* u_i *and* a_k (or, (u_i, a_k)-*contradicts* v, for short) if (6.1) is realizable for u_i but is not true for v. Our aim is to recognize for given objects $u_i \in U$ and $v \in \mathcal{U}(S)$, and given attribute a_k such that $a_k(u_i) \neq a_k(v)$ if there exists a rule from $Det_A(S)$ which (u_i, a_k)-contradicts v.

Let

$$M(u_i, v) = \{a_j : a_j \in A, a_j(u_i) = a_j(v)\} ,$$

and

$$P(u_i, v, a_k) = \{a_k(u) : u \in U, a_j(u) = a_j(v) \text{ for any } a_j \in M(u_i, v)\} .$$

Note that $|P(u_i, v, a_k)| \geq 1$.

Proposition 6.1. *Let* $S = (U, A)$ *be an information system,* $u_i \in U$, $v \in \mathcal{U}(S)$, $a_k \in A$ *and* $a_k(u_i) \neq a_k(v)$. *Then, in* $Det_A(S)$ *there exists a rule* (u_i, a_k)-*contradicting* v *if and only if* $|P(u_i, v, a_k)| = 1$.

Proof. Let $|P(u_i, v, a_k)| = 1$ and $P(u_i, v, a_k) = \{b\}$. In this case, the rule

$$\bigwedge_{a_j \in M(u_i,v)} (a_j(x) = a_j(v)) \rightarrow a_k(x) = b , \tag{6.2}$$

belongs to $Det_A(S)$, is realizable for u_i, and is not true for v, since $a_k(v) \neq a_k(u_i) = b$. Therefore, (6.2) is a rule from $Det_A(S)$ which (u_i, a_k)-contradicts v.

Let us assume that there exists a rule (6.1) from $Det_A(S)$, (u_i, a_k)-contradicting v. Since (6.1) is realizable for u_i and is not true for v, we have $a_{j_1}, \ldots, a_{j_t} \in M(u_i, v)$. Also (6.1) is true for S. Hence, the rule

$$\bigwedge_{a_j \in M(u_i,v)} (a_j(x) = a_j(v)) \rightarrow a_k(x) = b_k$$

is true for S. Therefore, $P(u_i, v, a_k) = \{b_k\}$ and $|P(u_i, v, a_k)| = 1$.   □

From Proposition 6.1 it follows that there exists polynomial algorithm recognizing, for a given information system $S = (U, A)$, given objects $u_i \in U$ and $v \in \mathcal{U}(S)$, and a given attribute $a_k \in A$ such that $a_k(u_i) \neq a_k(v)$, if there exists a rule from $Det_A(S)$, (u_i, a_k)-contradicting v.

This algorithm constructs the set $M(u_i, v)$ and the set $P(u_i, v, a_k)$. The considered rule exists if and only if $|P(u_i, v, a_k)| = 1$.

We also use the notion of *deterministic characteristic table* $D(S, v)$, where $v \in \mathcal{U}(S)$. This is a table with m columns and n rows. The entries of this table are binary (i.e., from $\{0, 1\}$). The number 0 is at the intersection of i-th row and

k-th column if and only if $a_k(u_i) \neq a_k(v)$ and there exists a rule from $Det_A(S)$, (u_i, a_k)-contradicting v.

From Proposition 6.1 it follows that there exists a polynomial algorithm which for a given information system $S = (U, A)$ and a given object $v \in \mathcal{U}(S)$ constructs the deterministic characteristic table $D(S, v)$.

6.1.3   Inhibitory Characteristic Tables

Let us consider a rule

$$(a_{j_1}(x) = b_1) \wedge \ldots \wedge (a_{j_t}(x) = b_t) \rightarrow a_k(x) \neq b_k , \tag{6.3}$$

where $t \geq 0$, $a_{j_1}, \ldots, a_{j_t}, a_k \in A$, $b_1,, \ldots, b_t, b_k \in \omega$, and numbers $j_1, \ldots, j_t, k$ are pairwise different. Such rules are called *inhibitory association* rules. Later we will omit the word "association". The rule (6.3) will be called *realizable for an object* $u \in \mathcal{U}(S)$ if $a_{j_1}(u) = b_1, \ldots, a_{j_t}(u) = b_t$. The rule (6.3) will be called *true for an object* $u \in \mathcal{U}(S)$ if $a_k(u) \neq b_k$ or (6.3) is not realizable for u. The rule (6.3) will be called *true for S* if it is true for any object from U. The rule (6.3) will be called *realizable for S* if it is realizable for at least one object from U. We denote by $Inh_A(S)$ the set of all inhibitory rules each of which is true for S and realizable for S.

Let $u_i \in U$, $v \in \mathcal{U}(S)$, $a_k \in A$ and $a_k(u_i) \neq a_k(v)$. We say that a rule (6.3) from $Inh_A(S)$ contradicts v *relative to the object* u_i *and the attribute* a_k (or (u_i, a_k)-*contradicts* v, for short) if (6.3) is realizable for u_i but is not true for v. Our aim is to recognize for given objects $u_i \in U$ and $v \in \mathcal{U}(S)$, and given attribute a_k such that $a_k(u_i) \neq a_k(v)$ if there exists a rule from $Inh_A(S)$, (u_i, a_k)-contradicting v.

Proposition 6.2. *Let* $S = (U, A)$ *be an information system,* $u_i \in U$, $v \in \mathcal{U}(S)$, $a_k \in A$ *and* $a_k(u_i) \neq a_k(v)$. *Then in* $Inh_A(S)$ *there is a rule* (u_i, a_k)-*contradicting* v *if and only if* $a_k(v) \notin P(u_i, v, a_k)$.

Proof. Let $a_k(v) \notin P(u_i, v, a_k)$. In this case, the rule

$$\bigwedge_{a_j \in M(u_i, v)} (a_j(x) = a_j(v)) \rightarrow a_k(x) \neq a_k(v) , \tag{6.4}$$

belongs to $Inh_A(S)$, is realizable for u_i, and is not true for v. Therefore, (6.4) is a rule from $Inh_A(S)$, (u_i, a_k)-contradicting v.

Let us assume that there exists a rule (6.3) from $Inh_A(S)$, (u_i, a_k)-contradicting v. In particular, it means that $a_k(v) = b_k$. Since (6.3) is realizable for u_i and is not true for v, we have $a_{j_1}, \ldots, a_{j_t} \in M(u_i, v)$. Since (6.3) is true for S, the rule

$$\bigwedge_{a_j \in M(u_i, v)} (a_j(x) = a_j(v)) \rightarrow a_k(x) \neq b_k$$

is true for S. Therefore, $a_k(v) \notin P(u_i, v, a_k)$. $\square$

From Proposition 6.2 it follows that there exists polynomial algorithm recognizing for a given information system $S = (U, A)$, given objects $u_i \in U$ and $v \in \mathcal{U}(S)$, and a given attribute $a_k \in A$ such that $a_k(u_i) \neq a_k(v)$ if there exists a rule from $Inh_A(S)$, (u_i, a_k)-contradicting v.

This algorithm constructs the set $M(u_i, v)$ and the set $P(u_i, v, a_k)$. The considered rule exists if and only if $a_k(v) \notin P(u_i, v, a_k)$.

In the sequel, we use the notion of *inhibitory characteristic table* $I(S, v)$, where $v \in \mathcal{U}(S)$. This is a table with m columns and n rows. The entries of this table are binary. The number 0 is at the intersection of i-th row and k-th column if and only if $a_k(u_i) \neq a_k(v)$ and there exists a rule from $Inh_A(S)$, (u_i, a_k)-contradicting v.

From Proposition 6.2 it follows that there exists a polynomial algorithm which for a given information system $S = (U, A)$ and a given object $v \in \mathcal{U}(S)$ constructs the inhibitory characteristic table $I(S, v)$.

6.1.4   Evaluation Functions

Let us denote by $\mathcal{T}$ the set of binary tables, i.e., tables with entries from $\{0, 1\}$ and let us consider a partial order $\preceq$ on $\mathcal{T}$. Let $Q_1, Q_2 \in \mathcal{T}$. Then $Q_1 \preceq Q_2$ if and only if $Q_1 = Q_2$ or Q_1 can be obtained from Q_2 by changing some entries from 1 to 0.

An *evaluation function* is an arbitrary function $W : \mathcal{T} \to [0, 1]$ such that $W(Q_1) \leq W(Q_2)$ for any $Q_1, Q_2 \in \mathcal{T}$, $Q_1 \preceq Q_2$. Let us consider five examples of evaluation functions W_1, W_2, W_3^α, W_4, and W_5^α, $0 < \alpha \leq 1$. Let Q be a table from $\mathcal{T}$ with m columns and n rows. Let $L_1(Q)$ be equal to the number of 1 in Q, $L_2(Q)$ be equal to the number of columns in Q filled by 1 only, $L_3^\alpha(Q)$ be equal to the number of columns in Q with at least $\alpha \times 100\%$ entries equal to 1, $L_4(Q)$ be equal to the number of rows in Q filled by 1 only, and $L_5^\alpha(Q)$ be equal to the number of rows in Q with at least $\alpha \times 100\%$ entries equal to 1. Then

$$W_1(Q) = \frac{L_1(Q)}{mn} \, , W_2(Q) = \frac{L_2(Q)}{m} \, , W_3^\alpha(Q) = \frac{L_3^\alpha(Q)}{m} \, ,$$

$$W_4(Q) = \frac{L_4(Q)}{n} \, , W_5^\alpha(Q) = \frac{L_5^\alpha(Q)}{n} \, .$$

It is clear that $W_2 = W_3^1$ and $W_4 = W_5^1$. Let $S = (U, A)$ be an information system and $v \in \mathcal{U}(S)$. Note that if $v \in U$, then for any α, $0 < \alpha \leq 1$, the following equalities hold: $W_1(D(S, v)) = W_2(D(S, v)) = W_3^\alpha(D(S, v)) = W_4(D(S, v)) = W_5^\alpha(D(S, v)) = 1$, $W_1(I(S, v)) = W_2(I(S, v)) = 1$, and $W_3^\alpha(I(S, v)) = W_4(I(S, v)) = W_5^\alpha(I(S, v)) = 1$.

6.2   Algorithms of Classification

A decision table T is a finite table filled by nonnegative integers. Each column of this table is labeled with a conditional attribute. Rows of the table are interpreted as tuples of values of conditional attributes on some objects. Each row is labeled with a nonnegative integer, which is interpreted as the value of decision attribute. Let T contain m columns labeled with conditional attributes $a_1, \ldots, a_m$. The set $\mathcal{U}(T) = \omega^m$ will be called the *universe* for the decision table T. For each object (tuple) $v \in \mathcal{U}(T)$, integers in v are interpreted as values of attributes $a_1, \ldots, a_m$ for this object.

We consider the following classification problem: for any object $v \in \mathcal{U}(T)$ it is required to generate a value of the decision attribute on v. To this end, we use DA-classification algorithm and IA-classification algorithm based on the deterministic characteristic table and the inhibitory characteristic table, respectively. DA-classification algorithm is a lazy learning algorithm based on deterministic association rules, and IA-classification algorithm is a lazy learning algorithm based on inhibitory association rules.

Let $Dec(T)$ be the set of values of decision attribute. For each $i \in Dec(T)$, let us denote by S_i the information system which tabular representation consists of all rows of T that are labeled with the decision i. Let W be an evaluation function.

DA-algorithm. For a given object v and $i \in Dec(T)$ we construct the deterministic characteristic table $D(S_i, v)$. Next, for each $i \in Dec(T)$ we find the value of the evaluation function W for $D(S_i, v)$. For each $i \in Dec(T)$ the value $W(D(S_i, v))$ is interpreted as the "degree" to which v belongs to S_i. As the value of decision attribute for v we choose $i \in Dec(T)$ such that $W(D(S_i, v))$ has the maximal value. If more than one such i exists, then we choose the minimal i for which $W(D(S_i, v))$ has the maximal value.

IA-algorithm. For a given object v and $i \in Dec(T)$ we construct the inhibitory characteristic table $I(S_i, v)$. Next, for each $i \in Dec(T)$ we find the value of the evaluation function W for $I(S_i, v)$. For each $i \in Dec(T)$ the value $W(I(S_i, v))$ is interpreted as the "degree" to which v belongs to S_i. As the value of decision attribute for v we choose $i \in Dec(T)$ such that $W(I(S_i, v))$ has the maximal value. If more than one such i exists, then we choose the minimal i for which $W(I(S_i, v))$ has the maximal value.

Example 6.3. Let us consider the decision table T (see Fig. 6.1) with two conditional attributes a_1 and a_2. The set $Dec(T)$ of values of the decision attribute is equal to $\{1, 2\}$. Hence, we work with two information systems S_1 and S_2 (see Fig. 6.1).

Let $v = (2, 2)$ be a new object. We now construct deterministic characteristic tables $D(S_1, v), D(S_2, v)$ and inhibitory characteristic tables $I(S_1, v), I(S_2, v)$ for this object and information systems S_1 and S_2 (see Fig. 6.2).

Let us consider outputs returned by DA-algorithm and IA-algorithm with evaluation functions W_1 and W_2 (definitions of these functions can be found at the end of the previous section).

DA-algorithm with function W_1. One can show that $W_1(D(S_1, v)) = 1$ and $W_1(D(S_2, v)) = 3/4$. Therefore, we assign to v the decision 1.

DA-algorithm with function W_2. One can show that $W_2(D(S_1, v)) = 1$ and $W_1(D(S_2, v)) = 0$. Therefore, we assign to v the decision 1.

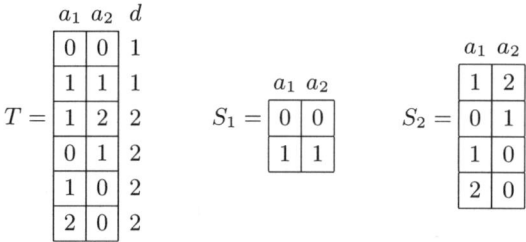

Fig. 6.1. Decision table T and information systems S_1, S_2

$$D(S_1, v) = \begin{array}{|c|c|} \hline 1 & 1 \\ \hline 1 & 1 \\ \hline \end{array} \quad D(S_2, v) = \begin{array}{|c|c|} \hline 0 & 1 \\ \hline 1 & 1 \\ \hline 1 & 1 \\ \hline 1 & 0 \\ \hline \end{array} \quad I(S_1, v) = \begin{array}{|c|c|} \hline 0 & 0 \\ \hline 0 & 0 \\ \hline \end{array} \quad I(S_2, v) = \begin{array}{|c|c|} \hline 0 & 1 \\ \hline 1 & 1 \\ \hline 1 & 1 \\ \hline 1 & 0 \\ \hline \end{array}$$

Fig. 6.2. Characteristic tables $D(S_1, v)$, $D(S_2, v)$, $I(S_1, v)$ and $I(S_2, v)$

IA-algorithm with function W_1. One can show that $W_1(I(S_1, v)) = 0$ and $W_1(I(S_2, v)) = 3/4$. Therefore, we assign to v the decision 2.

IA-algorithm with function W_2. One can show that $W_2(I(S_1, v)) = 0$ and $W_2(I(S_2, v)) = 0$. Therefore, we assign to v the decision 1.

6.3 Results of Experiments

We have performed experiments with the following algorithms: DA-algorithm with the evaluation functions W_1, W_2, W_3^α, W_4 and W_5^α, and IA-algorithm with the evaluation functions W_1, W_2, W_3^α, W_4 and W_5^α. To evaluate error rate of an algorithm on a decision table we use either train-and-test method or cross-validation method.

The following decision tables from [54] were used in our experiments: monk1 (7 attributes, 124 objects in training set, 432 objects in testing set), monk2 (7 attributes, 169 objects in training set, 432 objects in testing set), monk3 (7 attributes, 122 objects in training set, 432 objects in testing set), lymphography (19 attributes, 148 objects, 10-fold cross-validation), diabetes (9 attributes, 768 objects, 12-fold cross-validation, attributes are discretized by an algorithm from RSES2 [69]), breast-cancer (10 attributes, 286 objects, 10-fold cross-validation), primary-tumor (18 attributes, 339 objects, 10-fold cross-validation), balance-scale (5 attributes, 625 objects, 10-fold cross-validation), soybean-large (35 attributes, 307 objects in training set, 376 objects in testing set), post-operative (9 attributes, 90 objects, 10-fold cross-validation), hayes-roth (5 attributes, 132 objects in training set, 28 objects in testing set), lung-cancer (57 attributes, 32 objects, 10-fold cross-validation), solar-flare (13 attributes, 1066 objects in

Table 6.1. Results of experiments with evaluation functions W_1 and W_2

Decision table	DA, W_1	DA, W_2	IA, W_1	IA, W_2	err. rates [3]
monk1	0.292	0.443	0.114	0.496	0.000–0.240
monk2	0.260	0.311	0.255	0.341	0.000–0.430
monk3	0.267	0.325	0.119	0.322	0.000–0.160
lymphography	0.272	0.922	0.215	0.922	0.157–0.380
diabetes	0.348	0.421	0.320	0.455	0.224–0.335
breast-cancer	0.240	0.261	0.233	0.268	0.220–0.490
primary-tumor	0.634	0.840	0.634	0.846	0.550–0.790
average err. rate	0.330	0.503	0.270	0.521	0.164–0.404

training set, 323 objects in testing set). Missing values in decision tables are
filled by an algorithm from RSES2 [69]. We removed attributes of the kind
"name" that are distinct for each instance.

Table 6.1 contains results of experiments (error rates) for DA-algorithm and
IA-algorithm with the evaluation functions W_1 and W_2. The last row contains
average error rates. The last column contains some known results – the best and
the worst error rates for algorithms compared in the survey [3].

The obtained results show that the evaluation function W_1 is noticeably better
than the evaluation function W_2, and IA-algorithm with the evaluation function
W_1 is better than DA-algorithm with the evaluation function W_1. However, we
must note that for the evaluation function W_2 the results of DA-algorithm work
are, mainly, better than the results of IA-algorithm work.

The outputs returned by IA-algorithm with the evaluation function W_1 for
each of decision tables are comparable with the results reported in [3], but are
worse than the best results mentioned in [3].

Table 6.2 contains results of experiments (error rates) for two types of algorithms: DA-algorithm with the evaluation function W_3^α, and IA-algorithm
with the evaluation function W_3^α, where $\alpha \in \{0.50, 0.55, \ldots, 0.95, 1.00\}$. For algorithms of each type the best result (with the minimal error rate) and the
corresponding α to this result are presented in the table. The last row contains
average error rates. The last column contains some known results – the best and
the worst error rates for algorithms discussed in [3].

The obtained results show that the use of the parameterized evaluation func-
tion W_3^α, where $\alpha \in \{0.50, 0.55, \ldots, 0.95, 1.00\}$, makes it possible to improve the
performance of IA-algorithm with the evaluation function W_1 for tables monk3
and breast-cancer. With the exception of the table monk1, the results obtained
for IA-algorithm are better than the results obtained for DA-algorithm.

Table 6.3 contains results of experiments (error rates) for four types of algorithms: DA-algorithm with the evaluation functions W_5^α and W_4, and IA-
algorithm with the evaluation functions W_5^α and W_4, where $\alpha \in \{0.05, 0.10,$

Table 6.2. Results of experiments with evaluation function W_3^α

Decision table	DA, W_3^α	α	IA, W_3^α	α	err. rates [3]
monk1	0.172	0.95	0.195	0.85	0.000–0.240
monk2	0.301	0.95	0.283	0.95	0.000–0.430
monk3	0.325	1.00	0.044	0.65	0.000–0.160
lymphography	0.293	0.55	0.272	0.65	0.157–0.380
diabetes	0.421	1.00	0.351	0.95	0.224–0.335
breast-cancer	0.229	0.80	0.225	0.70	0.220–0.490
primary-tumor	0.658	0.75	0.655	0.70	0.550–0.790
average err. rate	0.343		0.289		0.164–0.404

Table 6.3. Results of experiments with evaluation functions W_4 and W_5^α

Decision table	DA, W_5^α	α	DA, W_4	IA, W_5^α	α	IA, W_4	err. rates [3]
monk1	0.285	0.60	0.288	0.109	0.90	0.109	0.000–0.240
monk2	0.237	0.90	0.237	0.227	0.90	0.227	0.000–0.430
monk3	0.218	0.55	0.315	0.109	0.85	0.109	0.000–0.160
lymphography	0.258	0.70	0.293	0.229	0.70	0.336	0.157–0.380
diabetes	0.331	0.70	0.365	0.320	0.85	0.335	0.224–0.335
breast-cancer	0.233	0.25	0.265	0.225	0.90	0.225	0.220–0.490
primary-tumor	0.667	0.95	0.667	0.658	0.90	0.679	0.550–0.790
average err. rate	0.319		0.348	0.269		0.289	0.164–0.404

..., 0.95, 1.00}. For the evaluation function W_5^α the best results (with the minimal error rates) and the corresponding α to these results are presented. The last row contains average error rates. The last column contains some known results – the best and the worst error rates for algorithms discussed in [3].

The obtained results show that the use of the evaluation function W_4 or the parameterized evaluation function W_5^α, where $\alpha \in \{0.05, 0.10, \ldots, 0.95, 1.00\}$, makes it possible to improve the performance of IA-algorithm with the evaluation function W_1 for tables monk1, monk2, monk3 and breast-cancer. With the exception of tables lymphography and primary-tumor (for the case of the evaluation function W_4), the results obtained for IA-algorithm are better than the results obtained for DA-algorithm.

In the last group of experiments, we are working with the following algorithms: DA-algorithm with the evaluation functions W_1, W_2, W_3^α, W_4 and W_5^α, and IA-algorithm with the evaluation functions W_1, W_2, W_3^α, W_4 and W_5^α, $\alpha \in \{0.05, 0.10, \ldots, 0.90, 0.95\}$. For evaluation functions W_3^α and W_5^α, we choose the minimal error rate among all considered α.

Table 6.4. Results of experiments with modified tables

Eval. func.	$D < I$	$I < D$	$D = I$
W_1	14	14	13
W_2	13	12	16
W_3^α	6	23	12
W_4	13	15	13
W_5^α	13	12	16
$\sum$	59	76	70

In our experiments, we use decision tables from [54]. For each original table, we choose a number of many-valued (with at least three values) attributes different from the decision attribute, and consider each such attribute as a new decision attribute. In this way, we obtain the same number of new decision tables as the number of chosen attributes.

The following decision tables from [54] were used in our experiments: balance-scale (4 new decision attributes), soybean-large (5 new decision attributes), post-operative (8 new decision attributes), hayes-roth (4 new decision attributes), lung-cancer (7 new decision attributes), solar-flare (7 new decision attributes). As a result we obtained 41 decision tables (original and new).

Table 6.4 includes the results of experiments. The rows of this table correspond to the evaluation functions W_1, W_2, W_3^α, W_4 and W_5^α. The column labeled with "$D < I$" contains the number of decision tables for which the error rate of DA-algorithm is less than the error rate of IA-algorithm. The column labeled with "$I < D$" contains the number of decision tables for which the error rate of IA-algorithm is less than the error rate of DA-algorithm. The column labeled with "$D = I$" contains the number of decision tables for which the error rate of DA-algorithm is equal to the error rate of IA-algorithm.

In particular (see Table 6.4), in 59 cases the error rate of DA-algorithm is less than the error rate of IA-algorithm, in 76 cases the error rate of IA-algorithm is less than the error rate of DA-algorithm, and in 70 cases DA-algorithm and IA-algorithm have the same error rate.

In experiments the DMES system [13] was used.

6.4 Conclusions

In this chapter, two families of lazy classification algorithms were investigated which are based on the evaluation of the number of types of true deterministic or inhibitory rules which give us "negative" information about new objects. Experimental results obtained for algorithms based on inhibitory rules are often better than the results for algorithms based on deterministic rules.

In 101 experiments the error rate of the classification algorithm based on inhibitory rules is less than the error rate of the classification algorithm based on deterministic rules. In 67 experiments the error rate of the algorithm based on inhibitory rules is greater than the error rate of the algorithm based on deterministic rules. In 72 experiments the algorithm based on inhibitory rules and the algorithm based on deterministic rules have the same error rates.

This fact can be considered as an experimental confirmation of theoretical results from Chap. 1 showing that the inhibitory rules provide a more powerful tool for representation of knowledge encoded in information systems than the deterministic rules. In the next chapter, we compare two similar lazy classification algorithms based on inhibitory decision rules and deterministic decision rules, respectively.

7 Lazy Classification Algorithms Based on Deterministic and Inhibitory Decision Rules

In this chapter, we consider the same classification problem as in Chaps. 5 and 6: for a given decision table T and a new object v it is required to generate a value of the decision attribute on v using values of conditional attributes on v.

We compare two lazy [1] classification algorithms based on deterministic and inhibitory decision rules of the forms

$$(a_1(x) = b_1) \wedge \ldots \wedge (a_t(x) = b_t) \rightarrow d(x) = b ,$$

$$(a_1(x) = b_1) \wedge \ldots \wedge (a_t(x) = b_t) \rightarrow d(x) \neq b$$

respectively, where $a_1, \ldots, a_t$ are conditional attributes, $b_1, \ldots, b_t$ are values of these attributes, d is the decision attribute and b is a value of d. By $Dec(T)$ we denote the set of values of the decision attribute d.

The first algorithm (DD-algorithm) was created and studied by J.G. Bazan [2, 3, 4]. This algorithm is based on deterministic decision rules. For a new object v and each decision $b \in Dec(T)$ we find (using polynomial-time algorithm) the number $D(T, b, v)$ of objects u from the decision table T such that there exists a deterministic decision rule r satisfying the following conditions: (i) r is true for the decision table T, (ii) r is realizable for u and v, and (iii) r has the equality $d(x) = b$ on the right-hand side. For the new object v we choose a decision $b \in Dec(T)$ for which the value $D(T, b, v)$ is maximal. Note that this approach was generalized by J.G Bazan [2, 3, 4] to the case of approximate decision rules, and by A. Wojna [88] to the case of decision tables with not only nominal but also numerical attributes.

The second algorithm (ID-algorithm) is based on inhibitory decision rules. For a new object v and each decision $b \in Dec(T)$ we find (using polynomial-time algorithm) the number $I(T, b, v)$ of objects u from the decision table T such that there exists an inhibitory decision rule r satisfying the following conditions: (i) r is true for the decision table T, (ii) r is realizable for u and v, and (iii) r has the relation $d(x) \neq b$ on the right-hand side. For the new object v we choose a decision $b \in Dec(T)$ for which the value $I(T, b, v)$ is minimal.

Results of experiments show that the algorithm based on inhibitory decision rules is often better than the algorithm based on deterministic decision rules.

P. Delimata et al.: Inhibitory Rules in Data Analysis, SCI 163, pp. 99–106.
springerlink.com   © Springer-Verlag Berlin Heidelberg 2009

This chapter is based on the paper [15].

The chapter consists of six sections. In Sect. 7.1, the notion of decision table is introduced. In Sects. 7.2 and 7.3, notions of deterministic and inhibitory decision rules are discussed. Section 7.4 contains definitions of two lazy classification algorithms. In Sect. 7.5, results of experiments are discussed. Section 7.6 contains short conclusions.

7.1 Decision Tables

In this section, we recall of some notions based on decision tables.

Let $T = (U, A, d)$ be a *decision table*, where $U = \{u_1, \ldots, u_n\}$ is a finite nonempty set of *objects*, $A = \{a_1, \ldots, a_m\}$ is a finite nonempty set of *conditional attributes* (functions defined on U), and d is the *decision attribute* (function defined on U). We assume that for each $u_i \in U$ and each $a_j \in A$ the value $a_j(u_i)$ and the value $d(u_i)$ belong to ω, where $\omega = \{0, 1, 2, \ldots\}$ is the set of nonnegative integers.

We also assume that the decision table $T = (U, A, d)$ is given by a *tabular representation*, i.e., a table with m columns and n rows. Columns of the table are labeled with attributes $a_1, \ldots, a_m$. At the intersection of i-th row and j-th column the value $a_j(u_i)$ is included. For $i = 1, \ldots, n$, the i-th row is labeled with the value $d(u_i)$ of the decision attribute d on the object u_i. For $i = 1, \ldots, n$, we identify the object $u_i \in U$ with the tuple $(a_1(u_i), \ldots, a_m(u_i))$, i.e., the i-th row of the tabular representation of the decision table T. By $Dec(T)$ we denote the set of values of the decision attribute d on objects from U.

The set $\mathcal{U}(T) = \omega^m$ is called the *universe* for the decision table T. Besides objects from U we consider also objects from $\mathcal{U}(T) \setminus U$. For any object (tuple) $v \in \mathcal{U}(T)$ and any attribute $a_j \in A$ the value $a_j(v)$ is equal to j-th integer in v.

7.2 Deterministic Decision Rules

Let us consider a rule

$$(a_{j_1}(x) = b_1) \wedge \ldots \wedge (a_{j_t}(x) = b_t) \rightarrow d(x) = b, \tag{7.1}$$

where $t \geq 0$, $a_{j_1}, \ldots, a_{j_t} \in A$, $b_1, \ldots, b_t \in \omega$, $b \in Dec(T)$ and numbers $j_1, \ldots, j_t$ are pairwise different. Such rules are called *deterministic decision* rules. The rule (7.1) is called *realizable for an object* $u \in \mathcal{U}(T)$ if $a_{j_1}(u) = b_1, \ldots, a_{j_t}(u) = b_t$ or $t = 0$. The rule (7.1) is called *true for an object* $u_i \in U$ if $d(u_i) = b$ or (7.1) is not realizable for u. The rule (7.1) is called *true for T* if it is true for any object from U. The rule (7.1) is called *realizable for T* if it is realizable for at least one object from U. By $Det_D(T)$ we denote the set of all deterministic decision rules which are true for T and realizable for T.

Our aim is to recognize for given objects $u_i \in U$ and $v \in \mathcal{U}(T)$, and given value $b \in Dec(T)$ if there exists a rule from $Det_D(T)$ which is realizable for u_i and v and has $d(x) = b$ on the right-hand side. Such a rule "supports" the assignment of the decision b to the new object v.

Let $M(u_i, v) = \{a_j : a_j \in A, a_j(u_i) = a_j(v)\}$ and $P(u_i, v) = \{d(u) : u \in U, a_j(u) = a_j(v)$ for any $a_j \in M(u_i, v)\}$. Note that if $M(u_i, v) = \emptyset$, then $P(u_i, v) = \{d(u) : u \in U\} = Dec(T)$.

Proposition 7.1. *Let* $T = (U, A, d)$ *be a decision table,* $u_i \in U$, $v \in \mathcal{U}(T)$, *and* $b \in Dec(T)$. *Then in* $Det_D(T)$ *there exists a rule, which is realizable for* u_i *and* v *and has* $d(x) = b$ *on the right-hand side, if and only if* $P(u_i, v) = \{b\}$.

Proof. Let $P(u_i, v) = \{b\}$. In this case, the rule

$$\bigwedge_{a_j \in M(u_i, v)} (a_j(x) = a_j(v)) \rightarrow d(x) = b \qquad (7.2)$$

belongs to $Det_D(T)$, is realizable for u_i and v, and has $d(x) = b$ on the right-hand side.

Let us assume that there exists a rule (7.1) from $Det_D(T)$, which is realizable for u_i and v, and has $d(x) = b$ on the right-hand side. Since (7.1) is realizable for u_i and v, we have $a_{j_1}, \ldots, a_{j_t} \in M(u_i, v)$. Since (7.1) is true for T, the rule (7.2) is true for T. Therefore, $P(u_i, v) = \{b\}$. $\qquad\qquad\square$

From Proposition 7.1 it follows that there exists a polynomial algorithm recognizing, for a given decision table $T = (U, A, d)$, given objects $u_i \in U$ and $v \in \mathcal{U}(T)$, and a given value $b \in Dec(T)$, if there exists a rule from $Det_D(T)$, which is realizable for u_i and v, and has $d(x) = b$ on the right-hand side. This algorithm constructs the set $M(u_i, v)$ and the set $P(u_i, v)$. The considered rule exists if and only if $P(u_i, v) = \{b\}$.

7.3 Inhibitory Decision Rules

Let us consider a rule

$$(a_{j_1}(x) = b_1) \wedge \ldots \wedge (a_{j_t}(x) = b_t) \rightarrow d(x) \neq b , \qquad (7.3)$$

where $t \geq 0$, $a_{j_1}, \ldots, a_{j_t} \in A$, $b_1,, \ldots, b_t \in \omega$, $b \in Dec(T)$, and numbers $j_1, \ldots, j_t$ are pairwise different. Such rules are called *inhibitory decision* rules. The rule (7.3) is called *realizable for an object* $u \in \mathcal{U}(T)$ if $a_{j_1}(u) = b_1, \ldots, a_{j_t}(u) = b_t$ or $t = 0$. The rule (7.3) is called *true for an object* $u_i \in U$ if $d(u_i) \neq b$ or (7.3) is not realizable for u_i. The rule (7.3) is called *true for* T if it is true for any object from U. The rule (7.3) is called *realizable for* T if it is realizable for at least one object from U. By $Inh_D(T)$ we denote the set of all inhibitory decision rules which are true for T and realizable for T.

Our aim is to recognize for given objects $u_i \in U$ and $v \in \mathcal{U}(T)$, and given value $b \in Dec(T)$ if there exists a rule from $Inh_D(T)$, which is realizable for u_i and v, and has $d(x) \neq b$ on the right-hand side. Such a rule "contradicts" the assignment of the decision b to the new object v.

Proposition 7.2. *Let* $T = (U, A, d)$ *be a decision table,* $u_i \in U$, $v \in \mathcal{U}(T)$, *and* $b \in Dec(T)$. *Then in* $Inh_D(T)$ *there exists a rule, which is realizable for* u_i *and* v, *and has* $d(x) \neq b$ *on the right-hand side, if and only if* $b \notin P(u_i, v)$.

Proof. Let $b \notin P(u_i, v)$. In this case, the rule

$$\bigwedge_{a_j \in M(u_i, v)} (a_j(x) = a_j(v)) \rightarrow d(x) \neq b \qquad (7.4)$$

belongs to $Inh_D(T)$, is realizable for u_i and v, and has $d(x) \neq b$ on the right-hand side.

Let us assume that there exists a rule (7.3) from $Inh_D(T)$, which is realizable for u_i and v, and has $d(x) \neq b$ on the right-hand side. Since (7.3) is realizable for u_i and v, we have $a_{j_1}, \ldots, a_{j_t} \in M(u_i, v)$. Since (7.3) is true for T, the rule (7.4) is true for T. Therefore, $b \notin P(u_i, v)$. $\square$

From Proposition 7.2 it follows that there exists a polynomial algorithm recognizing for a given decision table $T = (U, A, d)$, given objects $u_i \in U$ and $v \in \mathcal{U}(T)$, and a given value $b \in Dec(T)$ if there exists a rule from $Inh_D(T)$, which is realizable for u_i and v, and has $d(x) \neq b$ on the right-hand side. This algorithm constructs the set $M(u_i, v)$ and the set $P(u_i, v)$. The considered rule exists if and only if $b \notin P(u_i, v)$.

7.4 Algorithms of Classification

Let $T = (U, A, d)$ be a decision table. We consider the following classification problem: for an object $v \in \mathcal{U}(T)$ it is required to generate a value of the decision attribute d on v using only values of attributes from A on v. To this end, we use DD-classification algorithm (DD-algorithm) and ID-classification algorithm (ID-algorithm). DD-algorithm is a lazy learning algorithm based on deterministic decision rules. ID-algorithm is a lazy learning algorithm based on inhibitory decision rules.

DD-algorithm is based on the use of the parameter $D(T, b, v)$, $b \in Dec(T)$. This parameter is equal to the number of objects $u_i \in U$ for which there exists a rule from $Det_D(T)$, that is realizable for u_i and v, and has $d(x) = b$ on the right-hand side. From Proposition 7.1 it follows that there exists a polynomial algorithm which for a given decision table $T = (U, A, d)$, a given object $v \in \mathcal{U}(T)$ and a given value $b \in Dec(T)$ computes the value

$$D(T, b, v) = |\{u_i : u_i \in U, P(u_i, v) = \{b\}\}| .$$

DD-algorithm. For given object v and each $b \in Dec(T)$ we find the value of the parameter $D(T, b, v)$. As the value of the decision attribute for v we choose $b \in Dec(T)$ such that $D(T, b, v)$ has the maximal value. If more than one such b exists then we choose the minimal b for which $D(T, b, v)$ has the maximal value.

ID-algorithm is based on the use of the parameter $I(T, b, v)$, $b \in Dec(T)$. This parameter is equal to the number of objects $u_i \in U$ for which there exists a rule from $Inh_D(T)$, that is realizable for u_i and v, and has $d(x) \neq b$ on the right-hand side. From Proposition 7.2 it follows that there exists a polynomial algorithm which for a given decision table $T = (U, A, d)$, a given object $v \in \mathcal{U}(T)$ and a given value $b \in Dec(T)$ computes the value

$$I(T, b, v) = |\{u_i : u_i \in U, b \notin P(u_i, v)\}| \ .$$

ID-algorithm. For given object v and each $b \in Dec(T)$ we find the value of the parameter $I(T, b, v)$. As the value of the decision attribute for v we choose $b \in Dec(T)$ such that $I(T, b, v)$ has the minimal value. If more than one such b exists then we choose the minimal b for which $I(T, b, v)$ has the minimal value.

Example 7.3. Let us consider the decision table T with three conditional attributes a_1, a_2 and a_3, and five objects (rows) u_1, u_2, u_3, u_4 and u_5 (see Fig. 7.1). The set $Dec(T)$ of values of the decision attribute d is equal to $\{1, 2, 3\}$.

$$T = \begin{array}{c|c|c|c|c} & a_1 & a_2 & a_3 & d \\ \hline u_1 & 0 & 0 & 1 & 1 \\ u_2 & 1 & 0 & 0 & 1 \\ u_3 & 1 & 1 & 1 & 2 \\ u_4 & 0 & 1 & 0 & 3 \\ u_5 & 1 & 1 & 0 & 3 \end{array}$$

Fig. 7.1. Decision table T

Let $v = (0, 1, 1)$ be a new object. One can show that

$$M(u_1, v) = \{a_1, a_3\} \ , M(u_2, v) = \emptyset \ , M(u_3, v) = \{a_2, a_3\} \ ,$$
$$M(u_4, v) = \{a_1, a_2\} \ , M(u_5, v) = \{a_2\} \ ,$$

and

$$P(u_1, v) = \{1\} \ , P(u_2, v) = \{1, 2, 3\} \ , P(u_3, v) = \{2\} \ ,$$
$$P(u_4, v) = \{3\} \ , P(u_5, v) = \{2, 3\} \ .$$

Let us consider outputs returned by DD-algorithm and ID-algorithm.

DD-algorithm. One can show that $D(T, 1, v) = 1$, $D(T, 2, v) = 1$, and $D(T, 3, v) = 1$. Therefore, we assign to v the decision 1.

ID-algorithm. One can show that $I(T, 1, v) = 3$, $I(T, 2, v) = 2$, $I(T, 3, v) = 2$. Therefore, we assign to v the decision 2.

Remark 7.4. Let the decision attribute d of a decision table T has exactly two values b and c. Let v be a new object. One can show that $I(T, c, v) = D(T, b, v)$ and $I(T, b, v) = D(T, c, v)$. Therefore, DD-algorithm and ID-algorithm return for v the same output.

Table 7.1. Results of experiments with original decision tables

Decision table	DD-alg.	ID-alg.	Decision table	DD-alg.	ID-alg.
monk1	0.102	0.102	lenses	0.284	0.234
monk2	0.200	0.200	soybean-small	0.425	0.425
monk3	0.061	0.061	soybean-large	0.144	0.141
lymphography	0.215	0.208	zoo	0.149	0.060
diabetes	0.248	0.248	post-operative	0.345	0.356
breast-cancer	0.236	0.236	hayes-roth	0.072	0.143
primary-tumor	0.643	0.625	lung-cancer	0.600	0.600
balance-scale	0.213	0.234	solar-flare	0.022	0.022

Table 7.2. Results of experiments with modified decision tables

Decision table	New tables	DD opt	ID opt	DD average	ID average
lymphography	9	4	3	0.451	0.447
primary-tumor	3	1	2	0.352	0.343
balance-scale	4	0	4	0.805	0.761
soybean-large	5	0	3	0.138	0.136
zoo	1	0	1	0.270	0.260
post-operative	8	1	4	0.436	0.426
hayes-roth	4	1	3	0.545	0.501
lung-cancer	7	0	1	0.413	0.409
solar-flare	7	1	3	0.330	0.322

7.5 Results of Experiments

We have performed experiments with DD-algorithm and ID-algorithm. To evaluate error rate of an algorithm on a decision table we use either train-and-test method or cross-validation method.

The following decision tables from [54] were used in our experiments: monk1 (7 attributes, 124 objects in training set, 432 objects in testing set), monk2 (7 attributes, 169 objects in training set, 432 objects in testing set), monk3 (7 attributes, 122 objects in training set, 432 objects in testing set), lymphography (19 attributes, 148 objects, 10-fold cross-validation), diabetes (9 attributes, 768 objects, 12-fold cross-validation, attributes are discretized by an algorithm from RSES2 [69]), breast-cancer (10 attributes, 286 objects, 10-fold cross-validation), primary-tumor (18 attributes, 339 objects, 10-fold cross-validation), balance-scale (5 attributes, 625 objects, 10-fold cross-validation), lenses (5 attributes, 24 objects, 10-fold cross-validation), soybean-small (35 attributes, 47 objects, 10-fold cross-validation), soybean-large (35 attributes, 307 objects in training set, 376 objects in testing set), zoo (17 attributes, 101 objects, 10-fold cross-validation), post-operative (9 attributes, 90 objects, 10-fold cross-validation), hayes-roth (5 attributes, 132 objects in training set, 28 objects in testing set), lung-cancer (57 attributes, 32 objects, 10-fold cross-validation), solar-flare (13 attributes, 1066 objects in training set, 323 objects in testing set). Missing values in decision tables are filled by an algorithm from RSES2 [69]. We removed attributes of the kind "name" that are distinct for each instance.

Table 7.1 contains results of experiments (error rates) for DD-algorithm and ID-algorithm and original decision tables from [54].

For 3 decision tables the error rate of DD-algorithm is less than the error rate of ID-algorithm, for 5 decision tables the error rate of ID-algorithm is less than the error rate of DD-algorithm, and for 8 decision tables DD-algorithm and ID-algorithm have the same error rate.

Table 7.2 contains results of experiments for DD-algorithm and ID-algorithm and modified decision tables from [54]. For each original table from [54] we choose a number of many-valued (with at least three values) attributes different from the decision attribute, and consider each such attribute as new decision attribute. As a result, we obtain the same number of new decision tables as the number of chosen attributes (this number can be found in the column "New tables"). The column "DD opt" contains the number of new tables for which the error rate of DD-algorithm is less than the error rate of ID-algorithm. The column "ID opt" contains the number of new tables for which the error rate of ID-algorithm is less than the error rate of DD-algorithm. The column "DD average" contains the average error rate of DD-algorithm for new tables. The column "ID average" contains the average error rate of ID-algorithm for new tables. We choose many-valued attributes since for two-valued ones DD-algorithm and ID-algorithm have the same error rate for new decision tables (see Remark 7.4).

For 8 new decision tables the error rate of DD-algorithm is less than the error rate of ID-algorithm, for 24 new decision tables the error rate of ID-algorithm is less than the error rate of DD-algorithm, and for 16 new decision tables DD-algorithm and ID-algorithm have the same error rate. Note also that for each of the considered original tables the average error rate of ID-algorithm for new tables corresponding to the original one is less than the average error rate of DD-algorithm.

In experiments the DMES system [13] was used.

7.6 Conclusions

Results of experiments show that the algorithm based on inhibitory decision rules can be often better than the algorithm based on deterministic decision rules.

In 29 experiments the error rate of the classification algorithm based on inhibitory rules is less than the error rate of the classification algorithm based on deterministic rules. In 11 experiments the error rate of the algorithm based on inhibitory rules is greater than the error rate of the algorithm based on deterministic rules. In 24 experiments the algorithm based on inhibitory rules and the algorithm based on deterministic rules have the same error rates.

It means that classifiers can be constructed from inhibitory decision rules instead of deterministic decision rules.

There is an additional (intuitive) motivation for the use of inhibitory decision rules in classification algorithms: the inhibitory decision rules have much more chance to have larger support than the deterministic decision rules.

Final Remarks

In this monograph, we studied inhibitory decision and association rules. We showed that using inhibitory rules one can describe more knowledge encoded in information and decision systems than in the case of deterministic (standard) rules.

Unfortunately, for almost all k-valued information systems with the polynomial number of objects in the number of attributes the number of minimal (irreducible) inhibitory association rules is not polynomial in the number of attributes. In some sense analogous situation is with minimal inhibitory decision rules.

In such a situation, we can either use some heuristics for generating of relatively small sets of "important" inhibitory rules, or use lazy classification algorithms which in polynomial time can find an information about the whole set of true and realizable inhibitory rules for a given information or decision system.

We mainly studied greedy heuristics for construction of inhibitory rules. This choice is motivated by the following fact: under some natural assumptions on the class NP, greedy algorithms (with or without weights) are close to the best (from the point of view of precision) polynomial approximate algorithms for minimization of inhibitory rule complexity defined by the length or the total weight of attributes. We reported nontrivial lower bounds on the minimal complexity of inhibitory rules depending on an information received during the work of the greedy algorithm, and prove that for the most part of binary decision tables the greedy algorithm constructs short inhibitory rules with a relatively high accuracy.

We used greedy algorithms for construction of standard classifiers based on inhibitory and deterministic decision rules. Results of experiments with real-life decision tables from [54] showed that classifiers based on inhibitory rules are often more accurate than the classifiers based on deterministic rules.

We also constructed lazy classifiers based on inhibitory and deterministic association rules, and lazy classifiers based on inhibitory and deterministic decision rules. The results of experiments showed that the classifiers based on inhibitory rules are often better than the classifiers based on deterministic rules.

P. Delimata et al.: Inhibitory Rules in Data Analysis, SCI 163, pp. 107–108.
springerlink.com   © Springer-Verlag Berlin Heidelberg 2009

The obtained results will further to wider use of inhibitory rules in rough set theory and related theories such as test theory and logical analysis of data (LAD). The inhibitory rules play also an important role in the analysis and design of concurrent systems.

Note that an essential part of the results obtained in the monograph can be generalized to the case of information and decision systems with missing values [20, 27].

Finally, let us also observe that sets of inhibitory rules can be used to define some interesting frequent patterns [19, 21], e.g., frequent patterns satisfying constraints such as defined by bounds on deviation of decision over the set of objects satisfying such patterns [29]. In [29] there are presented heuristics based on genetic algorithms searching for such sets of inhibitory rules (generalized decision rules) and are also reported promising experimental results for classification of objects with application of such generalized rules.

References

1. Aha, D.W. (ed.): Lazy learning. Kluwer Academic Publishers, Dordrecht (1997)
2. Bazan, J.G.: Discovery of decision rules by matching new objects against data tables. In: Polkowski, L., Skowron, A. (eds.) RSCTC 1998. LNCS (LNAI), vol. 1424, Springer, Heidelberg (1998)
3. Bazan, J.G.: A comparison of dynamic and non-dynamic rough set methods for extracting laws from decision table. Rough sets in knowledge discovery 1. Methodology and applications (Studies in fuzziness and soft computing). Physica-Verlag, Heidelberg (1998)
4. Bazan, J.G.: Methods of approximate reasoning for synthesis of decision algorithms. Ph.D. Thesis, Warsaw University, Warsaw (in Polish) (1998)
5. Bazan, J.G., Nguyen, H.S., Nguyen, S.H., Synak, P., Wróblewski, J.: Rough set algorithms in classification problems. Rough set methods and applications: new developments in knowledge discovery in information systems (Studies in fuzziness and soft computing). Physica-Verlag, Heidelberg (2000)
6. Bazan, J.G., Skowron, A., Świniarski, R.: Rough sets and vague concept approximation: from sample approximation to adaptive learning. In: Peters, J.F., Skowron, A. (eds.) LNCS Transactions on Rough Sets V. LNCS, vol. 4100. Springer, Heidelberg (2006)
7. Boros, E., Hammer, P.L., Ibarki, T., Kogan, A., Mayoraz, E., Muchnik, I.: IEEE Transactions of Knowledge and Data Engineering 12, 292–306 (2000)
8. Bridewell, W., Langley, P., Todorovski, L., Dzeroski, S.: Machine Learning 71, 1–32 (2008)
9. Chegis, I.A., Yablonskii, S.V.: Trudy Matematicheskogo Instituta im. V.A. Steklova 51, 270–360 (1958) (in Russian)
10. Cheriyan, J., Ravi, R.: Lecture notes on approximation algorithms for network problems (1998), http://www.math.uwaterloo.ca/~jcheriya/lecnotes.html
11. Chvátal, V.: Mathematics of Operations Research 4, 233–235 (1979)
12. Crama, Y., Hammer, P.L., Ibaraki, T.: Ann. Oper. Res. 16, 299–326 (1988)
13. Data Mining Exploration System (DMES), http://www.univ.rzeszow.pl/rspn (Software)
14. Delimata, P., Moshkov, M.J., Skowron, A., Suraj, Z.: Two families of classification algorithms. In: An, A., Stefanowski, J., Ramanna, S., Butz, C.J., Pedrycz, W., Wang, W. (eds.) RSFDGrC 2007. LNCS (LNAI), vol. 4482, Springer, Heidelberg (2007)

15. Delimata, P., Moshkov, M.J., Skowron, A., Suraj, Z.: Comparison of lazy classification algorithms based on deterministic and inhibitory decision rules. In: Wang, G., Li, T., Grzymała-Busse, J.W., Miao, D., Skowron, A., Yao Y. (eds.) RSKT 2008. LNCS (LNAI), vol. 5009, Springer, Heidelberg (2008)

16. Delimata, P., Moshkov, M.J., Skowron, A., Suraj, Z.: Lazy classification algorithms based on deterministic and inhibitory rules. In: Proc. Information processing and management of uncertainty in knowledge-based systems (Malaga, Spain) (to appear, 2008)

17. Dmitriev, A.N., Zhuravlev, Y.I., Krendelev, F.P.: On mathematical principles for classification of objects and phenomena. In: Zhuravlev. Y.I., Makarov, S.V. (eds.) Diskretnyi Analiz 7. Akad. Nauk SSSR Sib. Otd. Inst. Mat., Novosibirsk (in Russian) (1966)

18. Feige, U.: A threshold of $\ln n$ for approximating set cover (preliminary version). In: Proc. 28th Annual ACM symposium on the theory of computing (Philadelphia, Pennsylvania, USA). ACM Press, New York (1996)

19. Frequent Itemset Mining Implementations Repository, http://fimi.cs.helsinki.fi/

20. Grzymała-Busse, J.W.: Data with missing attribute values: generalization of indiscernibility relation and rule induction. In: Peters, J.F., Skowron, A., Grzymała-Busse, J.W., Kostek, B., Świniarski, R.W., Szczuka, M.S. (eds.) LNCS Transactions on Rough Sets I. LNCS, vol. 3100. Springer, Heidelberg (2004)

21. Han, J., Pei, J.: Mining frequent pattern-growth: methodology and implications. SIGKDD Explor. Newsl. 2(2), 14–20 (2000)

22. Jankowski, A., Peters, J., Skowron, A., Stepaniuk, J.: Fundamenta Informaticae 85, 249-265 (2008)

23. Johnson, D.S.: J. Comput. System Sci. 9, 256–278 (1974)

24. Karp, R.M.: Reducibility among combinatorial problems. In: Miller, R.E., Thatcher, J.W. (eds.) Complexity of computer computations. Plenum Press, New York (1972)

25. Kearns, M.J.: The computational complexity of machine learning. MIT Press, Cambridge (1990)

26. Kodratoff, Y., Michalski, R.S. (eds.): Machine learning: an artificial intelligence approach, vol. III. Morgan Kaufmann, San Mateo, CA (1990)

27. Kryszkiewicz, M.: Information Sciences 113, 271–292 (1999)

28. Lovász, L.: Discrete Math. 13, 383–390 (1975)

29. Marszał-Paszek, B.: Belief functions in rough set theory with applications. Ph.D. Thesis, Institute of Computer Science PAS, Warsaw (in Polish) (2008) (submitted)

30. de Medeiros, A.K.A., Weijters, A.J.M.M., van der Aalst, W.M.P.: Data Mining and Knowledge Discovery 14, 245-304 (2007)

31. Michalski, R.S., Carbonell, T.J., Mitchell, T.M. (eds.): Machine learning: an artificial intelligence approach. TIOGA Publishing Co., Palo Alto, CA (1983)

32. Michalski, R.S., Carbonell, T.J., Mitchell, T.M. (eds.): Machine learning: an artificial intelligence approach, vol. II. Morgan Kaufmann, Los Altos, CA (1986)

33. Mitchel, T.M.: Machine Learning. McGraw-Hill Series in Computer Science, Boston, MA (1999)

34. Moshkov, M.J.: Electronic Notes in Theoretical Computer Science 82(4), 174–185 (2003)

35. Moshkov, M.J.: On greedy algorithm for partial cover construction. In: Lupanov, O.B. (ed.) Proc. Design and complexity of control systems (Nizhny Novgorod, Russia). Moscow University, Moscow (in Russian) (2003)

36. Moshkov, M.J.: On construction of the set of irreducible partial covers. In: Lupanov, O.B., Kasim-Zade, O.M., Chashkin, A.V., Steinhöfel, K. (eds.) SAGA 2005. LNCS, vol. 3777. Springer, Heidelberg (2005)
37. Moshkov, M.J.: On the set of partial reducts for the most part of binary decision tables. In: Czaja, L. (ed.) Proc. Concurrency, specification and programming 2 (Ruciane-Nida, Poland). Warsaw University, Warsaw (2005)
38. Moshkov, M.J.: Information Processing Letters 103, 66–70 (2007)
39. Moshkov, M.J., Piliszczuk, M.: On construction of partial reducts and bounds on their complexity. In: Wakulicz-Deja, A. (ed.) Proc. Decision support systems (Zakopane, Poland, 2004). University of Silesia, Katowice (2005)
40. Moshkov, M.J., Piliszczuk, M.: On partial tests and partial reducts for decision tables. In: Ślęzak, D., Wang, G., Szczuka, M.S., Düntsch, I., Yao, Y. (eds.) RSFDGrC 2005. LNCS (LNAI), vol. 3641. Springer, Heidelberg (2005)
41. Moshkov, M.J., Piliszczuk, M., Zielosko, B.: Greedy algorithm for construction of partial covers. In: Lupanov, O.B. (ed.) Proc. Problems of theoretical cybernetics (Penza, Russia). Moscow University, Moscow (in Russian) (2005)
42. Moshkov, M.J., Piliszczuk, M., Zielosko, B.: Lower bounds on minimal weight of partial reducts and partial decision rules. In: Wang, G., Peters, J.F., Skowron, A., Yao, Y. (eds.) RSKT 2006. LNCS (LNAI), vol. 4062, Springer, Heidelberg (2006)
43. Moshkov, M.J., Piliszczuk, M., Zielosko, B.: Lower bound on minimal weight of partial cover based on information about greedy algorithm work. In: Proc. Information processing and management of uncertainty in knowledge-based systems (Paris, France) (2006)
44. Moshkov, M.J., Piliszczuk, M., Zielosko, B.: On greedy algorithm with weights for construction of partial covers. In: Kłopotek, M.A., Wierzchoń, S.T., Trojanowski, K. (eds.) Proc. Intelligent information processing and web mining. Advances in soft computing. Springer, Heidelberg (2006)
45. Moshkov, M.J., Piliszczuk, M., Zielosko, B.: On partial covers, reducts and decision rules with weights. In: Peters, J.F., Skowron, A., Düntsch, I., Grzymała-Busse, J.W., Orłowska, E., Polkowski, L. (eds.) LNCS Transactions on Rough Sets VI. LNCS, vol. 4374. Springer, Heidelberg (2007)
46. Moshkov, M.J., Piliszczuk, M., Zielosko, B.: Fundamenta Informaticae 75, 357–374 (2007)
47. Moshkov, M.J., Piliszczuk, M., Zielosko, B.: On partial covers, reducts and decision rules. In: Peters, J.F., Skowron, A. (eds.) LNCS Transactions on Rough Sets VIII. LNCS, vol. 5084. Springer, Heidelberg (2008)
48. Moshkov, M.J., Piliszczuk, M., Zielosko, B.: Partial covers, reducts and decision rules in rough sets: theory and applications. Studies in Computational Intelligence, vol. 145. Springer, Heidelberg (to appear, 2008)
49. Moshkov, M.J., Skowron, A., Suraj, Z.: On testing membership to maximal consistent extensions of information systems. In: Greco, S., Hata, Y., Hirano, S., Inuiguchi, M., Miyamoto, S., Nguyen, H.S., Slowinski, R. (eds.) RSCTC 2006. LNCS (LNAI), vol. 4259, Springer, Heidelberg (2006)
50. Moshkov, M.J., Skowron, A., Suraj, Z.: On maximal consistent extensions of information systems. In: Wakulicz-Deja, A. (ed.) Proc. Decision support systems 1 (Zakopane, Poland, 2006). University of Silesia, Katowice (2007)
51. Moshkov, M.J., Skowron, A., Suraj, Z.: Fundamenta Informaticae 80(1-3), 247–258 (2007)
52. Moshkov, M.J., Skowron, A., Suraj, Z.: Information Sciences 178, 2600-2620 (2008)
53. Moshkov, M.J., Skowron, A., Suraj, Z.: On minimal inhibitory rules for almost all k-valued information systems, Journal of Applied Non-Classical Logics (submitted)

54. Newman, D.J., Hettich, S., Blake, C.L., Merz, C.J.: UCI repository of machine learning databases. University of California, Irvine (1998), http://www.ics.uci.edu/~mlearn/MLRepository.html

55. Nguyen, H.S.: Approximate Boolean reasoning: foundations and applications in data mining. In: Peters, J.F., Skowron, A. (eds.) LNCS Transactions on Rough Sets V. LNCS, vol. 4100. Springer, Heidelberg (2006)

56. Nguyen, H.S., Ślęzak, D.: Approximate reducts and association rules – correspondence and complexity results. In: Zhong, N., Skowron, A., Ohsuga, S. (eds.) Proc. RSFDGrC 1999. LNCS (LNAI), vol. 1711. Springer, Heidelberg (1999)

57. Nigmatullin, R.G.: The fastest descent method for covering problems. In: Proc. Questions of precision and efficiency of computer algorithms 5 (Kiev, USSR) (in Russian) (1969)

58. Pancerz, K.: An application of rough sets to the identification of concurrent system models. Ph.D. Thesis, Institute of Computer Science PAS, Warsaw (in Polish) (2006)

59. Pancerz, K., Suraj, Z.: Rough sets for discovering concurrent system models from data tables. In: Hassanien, A., Suraj, Z., Ślęzak, D., Lingras, P. (eds.) Rough computing: theories, technologies and applications. Idea Group, Inc. (2007)

60. Pawlak, Z.: Rough sets – theoretical aspects of reasoning about data. Kluwer Academic Publishers, Dordrecht (1991)

61. Pawlak, Z.: Bulletin of the EATCS 48, 178–190 (1992)

62. Pawlak. Z.: Rough set elements. In: Polkowski, L., Skowron, A. (eds.) Rough sets in knowledge discovery 1. Methodology and applications (Studies in fuzziness and soft computing). Physica-Verlag, Heidelberg (1998)

63. Pawlak, Z., Skowron, A.: Information Sciences 177, 3–27 (2007)

64. Pawlak, Z., Skowron, A.: Information Sciences 177, 28–40 (2007)

65. Pawlak, Z., Skowron, A.: Information Sciences 177, 41–73 (2007)

66. Piliszczuk, M.: On greedy algorithm for partial reduct construction. In: Czaja, L. (ed.) Proc. Concurrency, specification and programming 2 (Ruciane-Nida, Poland). Warsaw University, Warsaw (2005)

67. Quafafou, M.: Information Sciences 124, 301–316 (2000)

68. Raz, R., Safra, S.: A sub-constant error-probability low-degree test, and a sub-constant error-probability PCP characterization of NP. In: Proc. 29th Annual ACM symposium on the theory of computing (El Paso, Texas, USA). ACM Press, New York (1997)

69. Rough Set Exploration System (RSES), http://logic.mimuw.edu.pl/~rses

70. Rząsa, W., Suraj, Z.: A new method for determining of extensions and restrictions of information systems. In: Alpigini, J.J., Peters, J.F., Skowronek, J., Zhong, N. (eds.) RSCTC 2002. LNCS (LNAI), vol. 2475, Springer, Heidelberg (2002)

71. Skowron, A.: Rough sets in KDD. In: Shi, Z., Faltings, B., Musen, M. (eds.) Proc. 16th IFIP World computer congress (Beijing, China). Publishing House of Electronic Industry (2000)

72. Skowron, A., Rauszer, C.: The discernibility matrices and functions in information systems. In: Slowinski, R. (ed.) Intelligent decision support. Handbook of applications and advances of the rough set theory. Kluwer Academic Publishers, Dordrecht (1992)

73. Skowron, A., Suraj, Z.: Bulletin of the Polish Academy of Sciences 41(3), 237–254 (1993)

74. Skowron, A., Suraj, Z.: Discovery of concurrent data models from experimental tables: a rough set approach. In: Proc. Knowledge discovery and data mining (Montreal, Canada). AAAI Press, Menlo Park CA (1995)

75. Skowron, A., Świniarski, R., Synak, P.: Approximation spaces and information granulation. In: Peters, J.F., Skowron, A. (eds.) LNCS Transactions on Rough Sets III. LNCS, vol. 3400. Springer, Heidelberg (2005)
76. Slavík, P.: A tight analysis of the greedy algorithm for set cover (extended abstract). In: Proc. 28th Annual ACM symposium on the theory of computing (Philadelphia, Pennsylvania, USA). ACM Press, New York (1996)
77. Slavík, P.: Approximation algorithms for set cover and related problems. Ph.D. Thesis, University of New York at Buffalo (1998)
78. Ślęzak, D.: Approximate reducts in decision tables. In: Proc. Information processing and management of uncertainty in knowledge-based systems 3 (Granada, Spain) (1996)
79. Ślęzak, D.: Fundamenta Informaticae 44, 291–319 (2000)
80. Ślęzak, D.: Approximate decision reducts. Ph.D. Thesis, Warsaw University, Warsaw (in Polish) (2001)
81. Ślęzak, D.: Fundamenta Informaticae 53, 365–390 (2002)
82. Ślęzak, D., Wróblewski, J.: Order-based genetic algorithms for the search of approximate entropy reducts. In: Wang, G., Liu, Q., Yao, Y., Skowron, A. (eds.) RSFDGrC 2003. LNCS (LNAI), vol. 2639. Springer, Heidelberg (2003)
83. Soloviev, N.A.: Tests (theory, construction, applications). Nauka Publishers, Novosibirsk (in Russian) (1978)
84. Suraj, Z.: Rough set methods for the synthesis and analysis of concurrent processes. In: Polkowski, L., Tsumoto, S., Lin, T.Y. (eds.) Rough set methods and applications. New developments in knowledge discovery in information systems (Studies in fuzziness and soft computing). Physica-Verlag, Heidelberg (2000)
85. Suraj, Z.: Some remarks on extensions and restrictions of information systems. In: Ziarko, W., Yao, Y.Y. (eds.) RSCTC 2000. LNCS (LNAI), vol. 2005. Springer, Heidelberg (2001)
86. Suraj, Z., Pancerz, K.: A new method for computing partially consistent extensions of information systems: a rough set approach. In: Proc. Information processing and management of uncertainty in knowledge-based systems 3 (Paris, France). Editions EDK (2006)
87. Unnikrishnan, K.P., Ramakrishnan, N., Sastry, P.S., Uthurusamy, R.: 4th KDD Workshop on Temporal Data Mining: Network Reconstruction from Dynamic Data. The Twelfth ACM SIGKDD International Conference on Knowledge Discovery and Data (Philadelphia, USA). http://people.cs.vt.edu/~ramakris/kddtdm06/cfp.html (2006)
88. Wojna, A.: Analogy-based reasoning in classifier construction. In: Peters, J.F., Skowron, A. (eds.) LNCS Transactions on Rough Sets IV. LNCS, vol. 3700, Springer, Heidelberg (2005)
89. Wróblewski, J.: Fundamenta Informaticae 47, 351–360 (2001)
90. Yablonskii, S.V.: Tests. In: Glushkov, V.M. (ed.) Encyclopedia of cybernetics. Main Editorial Board of Ukrainian Soviet Encyclopedia, Kiev (in Russian) (1975)
91. Zhuravlev, Y.I.: Trudy Matematicheskogo Instituta im. V.A. Steklova 51, 143–157 (in Russian) (1958)
92. Zhuravlev, Y.I.: Inessential variables of partial Boolean functions. In: Vasil'ev, Y.L., Zhuravlev, Y.I., Korobkov, V.K., Krichevskii, R.E. (eds.) Diskretnyi Analiz 1. Akad. Nauk SSSR Sib. Otd. Inst. Mat., Novosibirsk (in Russian) (1963)
93. Ziarko, W.: Foundations of Computing and Decision Sciences 18, 381–396 (1993)
94. Zielosko, B.: On partial decision rules. In: Czaja, L. (ed.) Proc. Concurrency, specification and programming 2 (Ruciane-Nida, Poland). Warsaw University, Warsaw (2005)

Index